Investigative Strategies for Lead-Source Attribution at Superfund Sites Associated with Mining Activities

Committee on Sources of Lead Contamination at or near Superfund Sites

Board on Environmental Studies and Toxicology

Division on Earth and Life Studies

A Consensus Study Report of

The National Academies of

SCIENCES · ENGINEERING · MEDICINE

THE NATIONAL ACADEMIES PRESS

Washington, DC

www.nap.edu

THE NATIONAL ACADEMIES PRESS **500 Fifth Street, NW** **Washington, DC 20001**

This project was supported by Contract EP-C-14-005, TO#0015 between the National Academies of Sciences, Engineering, and Medicine and the US Environmental Protection Agency. Any opinions, findings, conclusions, or recommendations expressed in this publication are those of the authors and do not necessarily reflect the view of the organizations or agencies that provided support for this project.

International Standard Book Number-13: 978-0-309-46556-4
International Standard Book Number-10: 0-309-46556-7
Digital Object Identifier: https://doi.org/10.17226/24898

Additional copies of this publication are available for sale from the National Academies Press, 500 Fifth Street, NW, Keck 360, Washington, DC 20001; (800) 624-6242 or (202) 334-3313; http://www.nap.edu.

Printed in the United States of America

Suggested citation: National Academies of Sciences, Engineering, and Medicine. 2017. *Investigative Strategies for Lead-Source Attribution at Superfund Sites Associated with Mining Activities*. Washington, DC: The National Academies Press. doi: https://doi.org/10.17226/24898.

The National Academies of

SCIENCES · ENGINEERING · MEDICINE

The **National Academy of Sciences** was established in 1863 by an Act of Congress, signed by President Lincoln, as a private, nongovernmental institution to advise the nation on issues related to science and technology. Members are elected by their peers for outstanding contributions to research. Dr. Marcia McNutt is president.

The **National Academy of Engineering** was established in 1964 under the charter of the National Academy of Sciences to bring the practices of engineering to advising the nation. Members are elected by their peers for extraordinary contributions to engineering. Dr. C. D. Mote, Jr., is president.

The **National Academy of Medicine** (formerly the Institute of Medicine) was established in 1970 under the charter of the National Academy of Sciences to advise the nation on medical and health issues. Members are elected by their peers for distinguished contributions to medicine and health. Dr. Victor J. Dzau is president.

The three Academies work together as the **National Academies of Sciences, Engineering, and Medicine** to provide independent, objective analysis and advice to the nation and conduct other activities to solve complex problems and inform public policy decisions. The National Academies also encourage education and research, recognize outstanding contributions to knowledge, and increase public understanding in matters of science, engineering, and medicine.

Learn more about the National Academies of Sciences, Engineering, and Medicine at **www.nationalacademies.org.**

COMMITTEE ON SOURCES OF LEAD CONTAMINATION AT OR NEAR SUPERFUND SITES

Members

EDWARD J. BOUWER (*Chair*), Johns Hopkins University, Baltimore, MD
MARK D. BARTON, University of Arizona, Tucson, AZ
ERIC A. BETTERTON, University of Arizona, Tucson, AZ
JOEL D. BLUM, University of Michigan, Ann Arbor, MI
SUSAN L. BRANTLEY, Pennsylvania State University, University Park, PA
JUDITH C. CHOW, Desert Research Institute, Reno, NV
SCOTT E. FENDORF, Stanford University, Stanford, CA
ROBERT E. HAZEN, New Jersey Department of Environmental Protection, Trenton, NJ
CHRIS E. JOHNSON, Syracuse University, Syracuse, NY
WILLIAM I. MANTON, University of Texas at Dallas, Richardson, TX
JERRY R. MILLER, Western Carolina University, Cullowhee, NC
PEGGY A. O'DAY, University of California, Merced, CA
JOHN TOLL, Windward Environmental LLC, Seattle, WA
WARREN H. WHITE, University of California, Davis, CA

Staff

ELLEN K. MANTUS, Project Director
RAYMOND WASSEL, Scholar and Director of Environmental Studies
MIRSADA KARALIC-LONCAREVIC, Manager, Technical Information Center
RADIAH ROSE-CRAWFORD, Manager, Editorial Projects
TAMARA DAWSON, Program Coordinator

Sponsors

US ENVIRONMENTAL PROTECTION AGENCY

BOARD ON ENVIRONMENTAL STUDIES AND TOXICOLOGY

Members

WILLIAM H. FARLAND (*Chair*), Colorado State University, Fort Collins, CO
RICHARD A. BECKER, American Chemistry Council, Washington, DC
E. WILLIAM COLGLAZIER, American Association for the Advancement of Science, Washington, DC
DOMINIC M. DITORO, University of Delaware, Newark, DE
DAVID C. DORMAN, North Carolina State University, Raleigh, NC
CHARLES T. DRISCOLL, JR., Syracuse University, Syracuse, NY
ANNE FAIRBROTHER, Exponent Inc., Philomath, OR
GEORGE GRAY, The George Washington University, Washington, DC
STEVEN P. HAMBURG, Environmental Defense Fund, New York, NY
ROBERT A. HIATT, University of California, San Francisco, CA
SAMEUL KACEW, University of Ottawa, Ontario
H. SCOTT MATTHEWS, Carnegie Mellon University, Pittsburgh, PA
ROBERT PERCIASEPE, Center for Climate and Energy Solutions, Arlington, VA
R. CRAIG POSTLEWAITE, Burke, VA
MARK A. RATNER, Northwestern University, Evanston, IL
JOAN B. ROSE, Michigan State University, East Lansing, MI
GINA M. SOLOMON, California Environmental Protection Agency, Sacramento, CA
ROBERT M. SUSSMAN, Sussman and Associates, Washington, DC
DEBORAH L. SWACKHAMMER, University of Minnesota, St. Paul, MN
PETER S. THORNE, University of Iowa, Iowa City, IA

Senior Staff

TERESA A. FRYBERGER, Director
ELLEN K. MANTUS, Scholar and Director of Risk Assessment
RAYMOND A. WASSEL, Scholar and Director of Environmental Studies
SUSAN N. J. MARTEL, Senior Program Officer for Toxicology

Acknowledgments

This Consensus Study Report was reviewed in draft form by individuals chosen for their diverse perspectives and technical expertise. The purpose of this independent review is to provide candid and critical comments that will assist the National Academies of Sciences, Engineering, and Medicine in making each published report as sound as possible and to ensure that it meets the institutional standards for quality, objectivity, evidence, and responsiveness to the study charge. The review comments and draft manuscript remain confidential to protect the integrity of the deliberative process.

We thank the following individuals for their review of this report:

Teresa Bowers, Gradient
Edward Boyle, Massachusetts Institute of Technology
Gordon Brown, Stanford University
John Dilles, Oregon State University
Joseph Graney, Binghamton University, The State University of New York
Thomas Johnson, University of Illinois at Urbana-Champaign
Lin Ma, University of Texas at El Paso
Peter Nico, Lawrence Berkeley National Laboratory
Robert Seal, US Geological Survey
Aaron Thompson, University of Georgia

Although the reviewers listed above provided many constructive comments and suggestions, they were not asked to endorse the conclusions or recommendations of this report nor did they see the final draft before its release. The review of this report was overseen by David Dzombak, Carnegie Mellon University, and Michael Kavanaugh, Geosyntec Consultants. They were responsible for making certain that an independent examination of this report was carried out in accordance with the standards of the National Academies and that all review comments were carefully considered. Responsibility for the final content rests entirely with the authoring committee and the National Academies.

The committee gratefully acknowledges the following for their presentations to the committee during open sessions: Teresa Bowers, Gradient; Michele Burgess, US Environmental Protection Agency; Jonathan Garoutte, Missouri Department of Health and Senior Services; Art Hebrank, Missouri Mines State Historic Site; Cheryl Seeger, Missouri Geological Survey; Valerie Wilder, Missouri Department of Natural Resources; Emitt Witt, US Geological Survey; and Mark Yingling, The Doe Run Resources Corporation. The committee thanks Michele Burgess and Marc Stifelman, US Environmental Protection Agency, for providing background materials to the committee during the study process and especially Greg Bach, US Environmental Protection Agency, for organizing the committee's tour of the Southeast Missouri Lead Mining District. The committee is also grateful for the assistance of Norman Grossblatt who served as the report editor.

Contents

BOXES, FIGURES, AND TABLES

BOXES

FIGURES

TABLES

Summary

The Superfund program of the US Environmental Protection Agency (EPA) was created in the 1980s to address human-health and environmental risks posed by abandoned or uncontrolled hazardous-waste sites. Identification of Superfund sites and their remediation can be an expensive multistep process. As part of this process, EPA attempts to identify parties that are responsible for the contamination and thus financially responsible for remediation. Identification of potentially responsible parties is complicated because Superfund sites can have a long history of use and involve contaminants that can have many sources. Such is often the case for mining sites that involve metal contamination; metals occur naturally in the environment, they can be contaminants in the wastes generated at or released from the sites, and they can be used in consumer products, which can degrade and release the metals back to the environment. Lead as an environmental contaminant associated with mining sites provides a prime example of the complexities that are involved in identifying potential sources in that it meets all the criteria noted. Given the complexities surrounding lead contamination, Congress asked EPA to commission a study by the National Academies of Science, Engineering, and Medicine to determine whether there are techniques or methods that can help to distinguish sources of lead at Superfund sites associated with mining activities. Accordingly, the National Academies convened the Committee on Sources of Lead Contamination at or near Superfund Sites, which prepared this report.

THE COMMITTEE'S TASK AND THE SOUTHEAST MISSOURI LEAD MINING DISTRICT

In brief, the committee was asked to examine the extent to which various sources contribute to environmental lead contamination at Superfund sites that are near lead-mining areas and to focus on sources that contribute to lead contamination at sites near the Southeast Missouri Lead Mining District.[1] The committee interpreted its task as a request for investigative strategies that could provide the evidence needed to identify lead sources in various environmental media. It did not consider its task to be determining lead sources; such an activity would require extensive sampling, analysis, and interpretation that the committee was not equipped to perform. As directed, the committee focused its efforts on whether sources of lead in the Southeast Missouri Lead Mining District could be distinguished by using various investigative strategies. The committee emphasizes that it did not consider exposure pathways, which would need to be considered for identifying which lead sources are contributing to human exposure.

The Southeast Missouri Lead Mining District provides a prime example of the complexities that a site might pose. It is composed primarily of three large subdistricts—the Old Lead Belt, Mine La Motte-Fredericktown, and the Viburnum Trend—and has a long history of lead mining: it began in the early 1700s and continues to this day. Some areas in Southeast Missouri are contaminated with lead, and several have been designated as Superfund sites. Determining the sources of lead at the sites, however, is complicated. First, the area is rich in naturally occurring lead-ore deposits that are at or near the surface. Second, as noted, the area has a long history of mining activities. Third, an enormous amount of wastes known as chat or tailings has been produced as a result of the mining activities; some of the waste materials have been dispersed from their original disposal sites by natural forces, such as wind and rain, and some have been intentionally transported off site. Fourth, lead has been used extensively over centuries in various products. Thus, lead at a site could result from naturally occurring deposits, from historical mining activities, from current mining activities, from consumer products, or from some combination. The following sections describe potential sources and environmental dispersal mechanisms, highlight the commit-

[1]The full statement of task is in Chapter 1 of this report.

tee's investigative strategies and the implications for the Southeast Missouri Lead Mining District, and provide the committee's research recommendations.

SOURCES OF LEAD AND ENVIRONMENTAL DISPERSAL

Lead is a widespread, naturally occurring element that is found throughout the Earth's crust and occurs in virtually all rocks, soils, sediments, and waters at low concentrations. A variety of geologic processes can cause natural lead enrichment, and, in some cases, ore deposits are formed. Whether an ore deposit is mined depends on time, economics, technology, and natural processes. If various conditions are favorable, it might be decided to mine an ore deposit. Mining begins by physically extracting the ore; the ore is then processed to obtain a concentrate of lead minerals and smelted to obtain metallic lead, which is shipped to market for use in various consumer products (see Figure S-1). At each stage, materials are discarded, and this creates an opportunity for lead to be dispersed further in the environment. Historically, waste disposal was driven by what was easiest; thus, waste was generally discarded in the most efficient manner relative to the time and place. However, by the latter part of the 20th century, waste disposal was subject to regulatory controls and handled systematically in a mine's operating plan.

Sources of lead other than natural ones or those associated with mining can also cause environmental contamination. In the 20th century in the United States, lead was added to gasoline and paint; although those uses have been virtually eliminated, the legacy lead from them remains in soils and sediment and could be an important source of environmental contamination. Other uses of lead include plumbing materials, batteries, pesticides, agricultural amendments, ammunition, and other consumer products, such as jewelry, pottery, and glass.

Lead can be dispersed in the environment by natural processes and through intentional transport. A primary natural transport mechanism is via water. Although lead concentrations in water are generally lower than those of other dissolved species, lead can be transported in natural waters attached to colloidal or suspended particles. Depending on environmental conditions, sediment-associated lead in the river bank can be reintroduced into the river over both short and long periods. Depending on the prevailing weather, airborne transport has the greatest potential to disperse pollutants on regional and global scales because air is not confined by geographic barriers. Furthermore, transport rates of contaminants are often higher by air than by groundwater and surface waters. Intentional transport of mining materials and wastes has been an important mechanism for dispersal of lead in the environment. In the Southeast Missouri Lead Mining District, mine wastes (waste rock, tailings, and slags) were

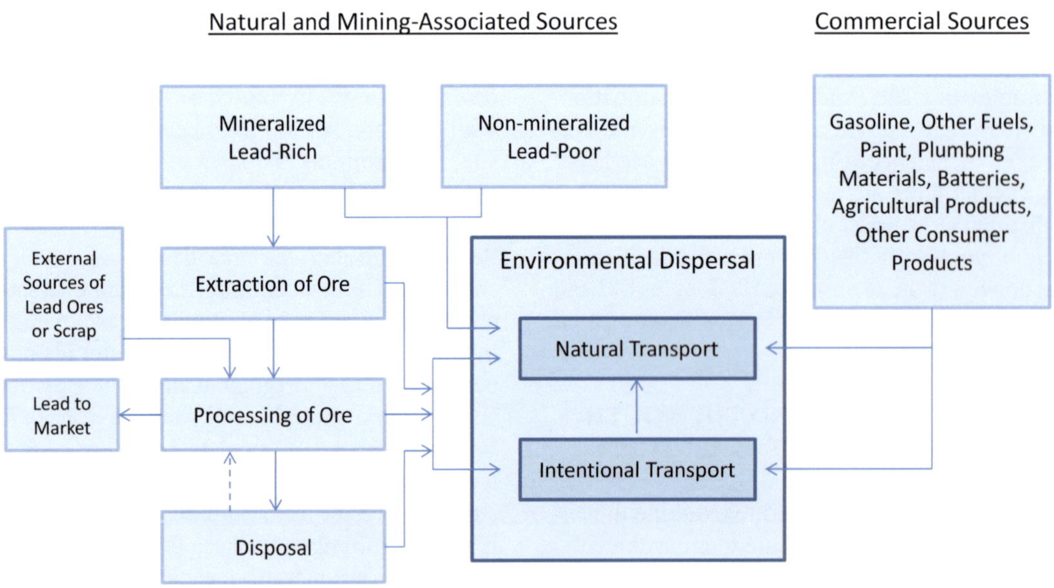

FIGURE S-1 Natural, mining-associated, and commercial sources can cause lead contamination in the environment through natural or intentional transport. *Natural transport* refers to environmental processes—such as wind, water flow, and soil erosion—that disperse materials; *intentional transport* refers to all human activities that move materials and have the potential for dispersal in the environment.

transported for use as construction fill, roadbed materials, agricultural amendments, and building materials.

INVESTIGATIVE STRATEGIES FOR LEAD-SOURCE ATTRIBUTION

Over the years, a wide array of analytic techniques have been used to determine the relative contribution of lead from natural and anthropogenic sources to soil, sediment, water, and air with varied success. Most techniques rely on a fingerprinting approach in which unique physical or chemical properties of the lead or lead-associated materials are used to distinguish sources from one another. The committee reviewed the techniques and provides investigative strategies for determining lead sources in soil, sediment, water, and air in Chapter 4 of this report. The approaches rely on systematic collection, analysis, and interpretation of multiple lines of evidence; an investigation continues until there is sufficient confidence that an accurate lead-source attribution has been reached. The committee emphasizes that its strategies are illustrative and that the sequence of steps will depend on site-specific conditions, availability of analytic techniques, and cost and time considerations.

The investigative strategies for soil, sediment, and water begin with a site characterization.[2] The goal here is to develop a site conceptual model that can guide investigations toward likely lead sources. Ideally, site characterization should document the source locations, the form of lead and other materials associated with sources, pathways and processes by which lead could be transported from a source, and, for anthropogenic sources, the history of lead use or processing. Once basic questions have been answered, a detailed sampling protocol can be developed for the specific site, and decisions can be made as to the best analytic approaches to pursue.

For the strategies noted, the committee concluded that lead isotopic analysis should be considered first for laboratory analysis of samples. Lead-isotope ratios in geologic materials vary owing to the production of lead isotopes from the radioactive decay of uranium (U) and thorium (Th). On geologic timescales, ^{235}U decays to ^{207}Pb, ^{238}U decays to ^{206}Pb, and ^{232}Th decays to ^{208}Pb; ^{204}Pb is primordial and has no radioactive parent. Various factors, such as geologic conditions and ages, result in a wide array of lead-isotope ratios among lead-bearing ore deposits and other geologic materials. Consequently, ore deposits can have unique lead-isotope ratios (signatures) that distinguish them from each other. When lead is purified from the ores and used in industry, the lead-containing commercial products retain the lead-isotope ratios of the ore deposits used to make the products. Thus, the technique works well if there is an isotopic contrast between the sources that one wishes to distinguish, but it can be equivocal if there are overlapping isotopic ratios in the various potential sources. In most cases, some potential sources can be eliminated and thus focus the investigation on fewer possible sources.

If the isotopic analysis does not yield unique signatures or if additional confidence is needed to reach conclusions, the committee suggests other analytic techniques and methods, such as ones that rely on analysis of mineralogy, particle size, chemical speciation, or other source-associated elemental ratios or isotopes. Each method has the potential to provide additional evidence to identify lead sources. However, the committee emphasizes that even after extensive sampling and analysis it still might not be possible to attribute lead to particular sources with confidence.

IMPLICATIONS FOR THE SOUTHEAST MISSOURI LEAD MINING DISTRICT

Isotopic analytic techniques might be especially promising for distinguishing non-native ores (or products made from non-native ores) from the natural or locally mined materials in Southeast Missouri. The committee tested the possibility by compiling and comparing $^{206}Pb/^{207}Pb$ ratios of North American lead ores, gasoline, and paints from around the world. As noted, various ore deposits can have different lead-isotope ratios, and gasoline producers and paint manufacturers used lead mined from different ore bodies. Therefore, it is expected that in some specific cases the lead isotopic composition of gasoline and paint might differ from that of local ore bodies; in other cases, they might be the same and therefore not useful for source attribution. The lead ores from Southeast Missouri have an extremely unusual lead isotopic composition, and the isotopic composition of lead mined from Southeast Missouri (and adjacent areas in the tristate region) appears to be distinct from that of nearly all other major lead mines (and products made with lead from them) in North America.

Figure S-2 clearly shows that Southeast Missouri ores, especially those from the Old Lead Belt, have higher $^{206}Pb/^{207}Pb$ ratios than other ore deposits, gasoline, and paints used in the United States. The only exceptions are the lead ores from the Upper Mississippi Valley District, which were probably not produced in substantial quantities in the 20th century compared with those from Southeast Missouri, and a single sample from a house within 0.5 miles of the Herculaneum lead smelter that could be the result of lead contamination from the smelter. Thus, the committee concludes that in the case of the Southeast Missouri Lead Mining District, it is likely that lead-isotope ratios can be used to attribute the sources of lead in

[2]The investigative approach for air is different from that for the other environmental media in that it uses a tiered approach that focuses primarily on modeling for source attribution, although similar techniques to analyze airborne lead particles are noted.

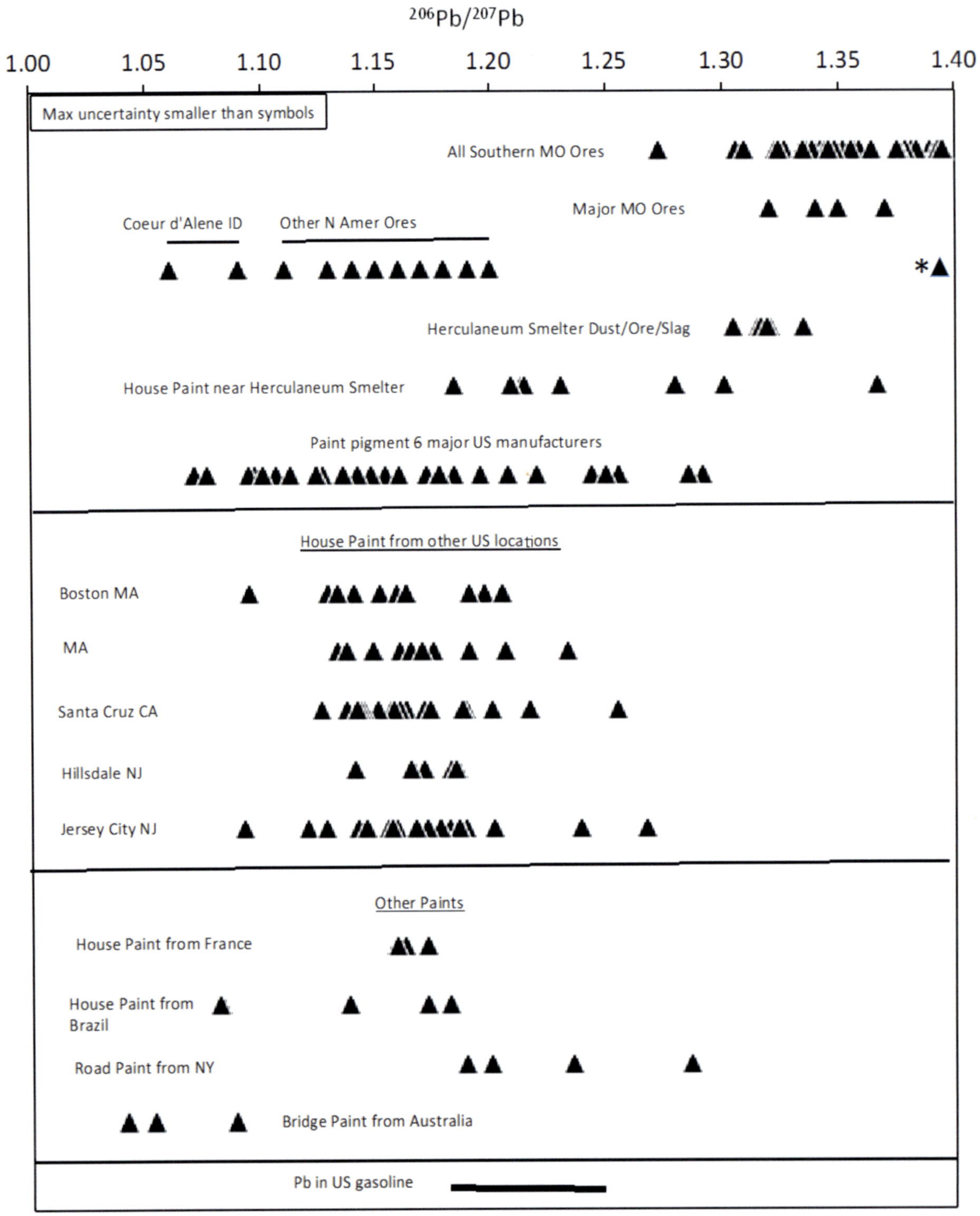

FIGURE S-2 Compilation of $^{206}Pb/^{207}Pb$ ratios in North American ore deposits, gasoline, and paint samples. The asterisked triangle represents the average $^{206}Pb/^{207}Pb$ ratios in ores from the Upper Mississippi Valley District. References for the data sources can be found in Chapter 5, Figure 5-1, and corresponding reference list.

environmental media between mining materials and gasoline and paint in most cases. If there are non-native ores or other lead-containing consumer products at the site—such as plumbing materials, batteries, or pesticides—lead-isotope studies of those materials might establish their potential role as a source of lead contamination at a site.

Distinguishing lead from natural (background) sources and lead from local mining materials or waste is much more challenging in contaminated soil. Natural and mined materials from the same region will typically contain lead that exhibits similar lead isotopic composition, and this would hinder the use of lead isotopic analysis to determine source. However, one might be able to use other methods alone or in combination—such as analysis of grain size, structure, and composition of soil particles, and depth distributions of metals in soils—to attribute potential sources. For example, smelter-derived particles in soils often contain spherical grains that are composed of sulfides, oxides, or glassy particles, and their spatial distribution is also likely to vary as a function of wind direction and distance from the smelter. With time, however, weathering of minerals derived from primary ore or smelter sources in soils might alter or dissolve them and potentially disperse them in the environment. The committee emphasizes that the selection of specific properties to be used for source attribution must be based on physical, compositional, mineralogic, or geochemical differences in the materials that make up the various source materials that need to be distinguished. Furthermore, it might be necessary to use multiple methods to differentiate between sources. In some cases, even with multiple analytic approaches, source identification might not be definitive in distinguishing natural materials from local mined materials or wastes, especially in soils and sediments.

RESEARCH RECOMMENDATIONS

Additional research will provide opportunities to enhance the analytic techniques and methods that make up the committee's proposed investigative strategies and to reduce uncertainty in their implementation. The committee finds that characterizing potential sources has high priority for research and lists below several topics that deserve attention. The research needs are not listed in order of priority.

- *Improved characterization of lead isotopic composition of historical paints.* To evaluate the utility of lead isotopic analysis for distinguishing lead sources, additional study of lead isotopes will be necessary. There appear to be few data on the lead isotopic composition of house paint in Southeast Missouri, and the committee recommends systematic sampling and lead isotopic analysis of paint in houses built before 1980 throughout the Southeast Missouri Lead Mining District. Such study would give a much clearer indication of the lead isotopic composition of paint used in the area and would clarify the degree to which lead isotopes provide a method for attributing some lead sources.
- *Improved characterization of lead-bearing geologic materials and mining wastes.* Because lead isotopic analysis will probably not allow distinguishing natural materials from mining materials or wastes, physical and chemical properties of the latter materials need to be characterized better so that the properties can be used to help to distinguish sources. The goal is to identify fine-scale variability in the sources on the basis of isotopic composition, elemental chemistry, mineralogy, grain size, or texture.
- *Capability to model metal concentrations in soils as a function of depth.* The science of modeling sediment distributions in river floodplains is relatively advanced. In contrast, modeling of metal movement through soils is less developed. Studies could be completed on soils near known metal-emitting sources to map and model how the metals have moved, if at all, through the soils as a function of time.
- *Characterization of size distribution of suspended particles in air samples.* Tailings impoundments, chat piles, and smelter emissions are important sources of airborne lead in a mining district, and little is known about the size of suspended particles and their mobility. Size-selective sampling of air samples in the Southeast Missouri Lead Mining District should be conducted to determine the particle-size distribution. Size-selective sampling can be conducted upwind or downwind of a source, such as a smelter or tailings impoundment, to evaluate contributions from the source.
- *Characterization of source strength of airborne particles from mining wastes.* Little is known about rates of emission of airborne lead from stockpiles of mining wastes. Portable wind tunnels can be used to estimate the airborne particle emission rates and dust production from stockpiles of mining wastes and to measure the threshold wind velocity at a specific site that results in particle suspension and later transport downwind. That information can help to establish whether airborne lead from mining wastes is an important source of lead contamination downwind.
- *Investigation of industrial archeology as a tool for source identification.* Industrial archeology is the systematic study of material evidence associated with historical industrial activity. Although historical industrial processes were generally on a smaller scale than modern industry, they operated less efficiently, were more likely to use processes and materials that produced hazardous byproducts, and lacked pollution controls. An understanding of historical practices could shed light on their relative

importance as a source of legacy pollution that persists long after the industries have disappeared.

Although the committee was not asked to identify possible exposure pathways or methods for determining which lead source at a contaminated site might be the most important for human exposure, it did consider whether its investigative approaches could help to answer such questions. Blood lead is often used as a measure to determine whether (and how much) lead exposure has occurred. The committee's recommendation for using lead isotopic analysis also applies to investigating sources of blood lead. Although additional study would be needed to determine where and under what conditions lead isotopic analysis could provide unambiguous lead-source attribution for people who have high blood lead concentrations, the approach appears promising for the Southeast Missouri Lead Mining District and possibly other mining districts, such as Coeur d'Alene, Idaho.

1

Introduction

In 1980, Congress passed the Comprehensive Environmental Response, Compensation, and Liability Act (CERCLA), which established the Superfund program to address environmental contamination in the United States. As part of the Superfund program, the US Environmental Protection Agency (EPA) conducts site investigations, determines whether a site needs to be remediated, and attempts to identify parties who are responsible for the contamination and thus financially responsible for the remediation activities. Identifying potentially responsible parties is often difficult because sites can have a long history of use and involve contaminants that have many potential sources. That is often the case for sites that involve metal contamination; metals occur naturally in the environment, they can be contaminants in the wastes generated at or released from a site, and they can be used in consumer products, which can degrade and release the metals into the environment. Lead as an environmental contaminant associated with sites at or near lead-mining districts provides a primary example of the complexities that are involved in identifying potential sources. Lead occurs naturally in the environment, is present in wastes associated with lead-mining districts, and has been used in various consumer products that are known to have resulted in environmental contamination, such as leaded gasoline and paint. Given the complexities surrounding lead contamination, Congress asked EPA to commission a study by the National Academies of Sciences, Engineering, and Medicine to examine the extent to which various sources contribute to environmental lead contamination at Superfund sites that are near lead-mining areas. Accordingly, the National Academies convened the Committee on Sources of Lead Contamination at or near Superfund Sites, which prepared this report.

SUPERFUND PROGRAM

As noted, the Superfund program was established to address human-health and environmental risks posed by abandoned or uncontrolled hazardous-waste sites. CERCLA gave EPA the resources and authority to clean up sites and seek reimbursement from potentially responsible parties. The Superfund process begins with the identification or notification of a potentially hazardous release and involves several steps, which are described in Box 1-1. If EPA's investigation results in listing of a site on the National Priorities List (NPL), the site is eligible for cleanup under the Superfund program. As of March 2017, 1,337 sites were listed on the NPL (EPA 2017a). They include manufacturing plants, processing facilities, landfills, and mining sites and involve a wide array of contaminants. Some of the most complex sites are mining sites not only because of their size but because their contaminants can have multiple possible natural and anthropogenic sources; this is problematic when one is trying to identify contaminant sources and assign financial responsibilities for remediation activities. Appendix A provides information on the challenges of obtaining information that can help to determine financial responsibilities.

SOUTHEAST MISSOURI LEAD MINING DISTRICT

As noted, mining sites are typically large and complex, and the Southeast Missouri Lead Mining District provides a prime example of the complexities that a site might pose. It is composed of three large subdistricts—the Old Lead Belt, Mine La Motte–Fredericktown, and the Viburnum Trend—and several smaller ones (see Figure 1-1; Seeger 2008).[1] The area has a long history of lead mining that began in the early 1700s and continues today (Seeger 2008). Until the 1860s, the majority of mining was accomplished by digging shallow pits and extracting the ore near the surface (Buckley 1909; Seeger 2008). Mining practices changed dramatically in the 1860s; the Saint Joseph Lead Company was formed in 1864, and the diamond drill bit was introduced in 1869 (Bonne Terre

[1]The committee notes that considerable lead was also produced in the Washington County barite district in the same region.

News 2008; Mugel 2017). Those events led to extensive underground mining in the area and the development of a complex of mines in the Old Lead Belt that were 200–300 feet deep and connected by over 250 miles of underground railroad (Seeger 2008). Mines in the Old Lead Belt were finally closed by the early 1970s because the higher-grade ore had been exhausted. At that time, the large and recently discovered deposits of the Viburnum Trend were coming into production (MO DNR 2017). Today, the only active mining in the area is along the Viburnum Trend, from which Missouri continues to supply most of the US primary lead with byproduct zinc, copper, and silver (Guberman 2014).

The mining sites typically included milling operations in which the ore was processed and the lead-bearing minerals were extracted. There were also smelting operations in the area that converted the lead from its mineral form to its elemental state. Waste materials were generated in processing the ore, and in the early years a density separation process was used and generated a coarse waste material often referred to as chat that was accumulated in enormous piles at the milling sites (ATSDR 1996, 1998). In the later years, a finer grinding of the ore allowed a more efficient separation process (flotation) to extract the lead minerals while generating much finer waste mate-

BOX 1-1 Steps in the Superfund Assessment Process

Site Assessment: The first phase of the Superfund process includes the *preliminary assessment* and *site investigation* in which EPA gathers the site history and analyzes air, water, and soil samples to determine the risk that the site might pose to people or the environment. The data collected al-low EPA to determine whether any immediate action should be taken to protect public health or the environment and to calculate a score by using its hazard ranking system (HRS), which helps to determine whether the site should be listed on the NPL.

NPL Listing Process: This process is started if the HRS score indicates that the site should be included on the NPL and includes an announcement and public-comment period regarding a potential listing of the site on the NPL. Once the comments have been reviewed and evaluated, EPA determines whether the site should be listed. The NPL contains the worst hazardous-waste sites that require extensive cleanup measures.

Site Characterization: This stage is undertaken if a site is listed on the NPL. It includes a *remedial investigation* in which numerous samples of air, surface water, groundwater, sediment, soil, waste, and biota are taken and analyzed to determine the specific risks associated with the site. A *feasibility study* outlines possible options for cleanup and discusses costs, advantages, and disadvantages of each option. A proposed plan for cleanup is developed and announced, and public comment is sought.

Record of Decision: A record of decision contains the final decision regarding cleanup measures. It provides information on "site history, site description, site characteristics, community participation, enforcement activities, past and present activities, contaminated media, the contaminants present, description of response actions to be taken…the remedy selected for cleanup…[and] consideration of how the site could be used in the future."

Remedial Design and Remedial Action: The next phase of the process includes the development of specific engineering details concerning cleanup measures and the process of undertaking cleanup.

Construction and Post-construction Completion: Construction completion is the signal that all physical cleanup activities have been completed. Post-construction activities can be initiated, such as site monitoring that is required to ensure public and environmental safety.

NPL Deletion: The site is considered for deletion from the NPL when cleanup is considered complete and the site is no longer judged to pose a risk to people or the environment. An announcement is made concerning the delisting, and public comment is sought. After reviewing and evaluating the comments, EPA determines whether the site should be removed from the NPL.

Reuse: The goal of the Superfund process is restoration of a site to productive industrial, commercial, residential, or recreational use.

Source: EPA (2017b).

rials, known as tailings, that were pumped into settling ponds or impoundments (ATSDR 1996, 1998). The committee notes that mining waste or residual material from processing is typically referred to as tailings; however, the term *chat* is also used in the Southeast Missouri Lead Mining District to distinguish between coarse and fine waste material. Mining operations are discussed further in Chapter 2.

Some areas in Southeast Missouri are contaminated with lead, and several areas in St. Francois, Madison, Jefferson, and Washington Counties have been designated as Superfund sites (EPA 2017c). Determining the sources of lead at the sites, however, is complex. First, the area is rich in naturally occurring lead-ore deposits that are at or near the surface. Second, the area has a long history of mining activities. Third, an enormous amount of waste has been produced as a result of all the mining activities. For example, an estimated 250 million tons of tailings were generated from mining activities in the Old Lead Belt (ATSDR 1996, 1998). Some of the tailings have been redistributed from their original disposal sites by natural forces, such as wind and rain, and there has been intentional transport of

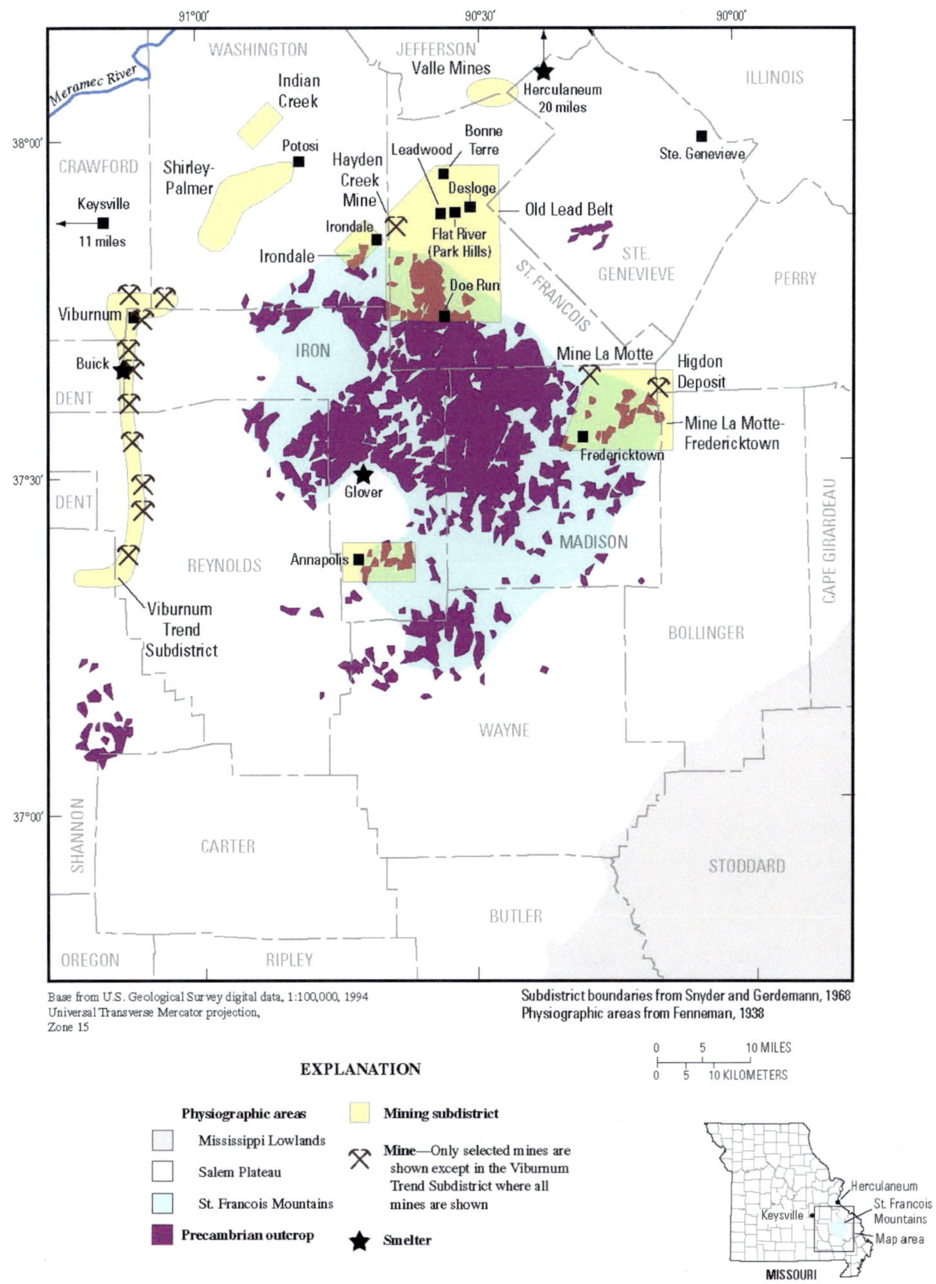

FIGURE 1-1 Southeast Missouri Lead Mining District. Source: Seeger (2008).

chat and other byproducts offsite (ATSDR 1996, 1998). Specifically, the tailings were sold as agricultural lime, used as fill material, incorporated into asphalt, and used to treat roads (Meyers 1989; ATSDR 1996). There are even anecdotal accounts of the tailings being used for children's sandboxes (Lee 2006). Fourth, the picture is complicated by the long history of use of lead in consumer products, such as paint and gasoline. Thus, lead at a site could result from naturally occurring deposits, from historical mining activities, from current mining activities, from consumer products, or from some combination, and identifying the sources becomes problematic.

THE COMMITTEE AND ITS TASK

The committee that was convened as a result of the agency's request included experts in aqueous and soil geochemistry, economic geology, environmental fate and transport, lead-source characterization, environmental data analysis, risk assessment, and Superfund policy. Appendix B provides biographic information on the committee members. As noted, the committee was asked to examine the extent to which various sources contribute to environmental lead contamination at Superfund sites that are situated near lead-mining areas and to focus specifically on sources contributing to lead contamination at sites near the Southeast Missouri Lead Mining District. The committee's verbatim statement of task is provided in Box 1-2; it is important to note that the committee was not asked to identify human exposure pathways.

THE COMMITTEE'S APPROACH TO ITS TASK

To accomplish its task, the committee held five meetings. During its first two meetings, the committee held open sessions to hear from the sponsor, mining-industry representatives, and regional scientists from the US Geological Survey, the Missouri Geological Survey, the Missouri Department of Natural Resources, and the Missouri Department of Health and Senior Services. As part of the second meeting, the committee toured portions of the Southeast Missouri Lead Mining District to gain a better appreciation of the natural geology and geography, the mining and tailings sites, and the remediation activities. Appendix C contains the agendas of the open sessions. The committee reviewed numerous scientific publications and all materials submitted to it from outside parties.

The committee interpreted its task as a request for investigative strategies that could provide the evidence needed to identify lead sources in various environmental media. It did not consider its task to be determining lead sources; such an activity would require extensive sampling, analysis, and interpretation that the committee was not equipped to perform. As directed, the committee focused its efforts on whether sources of lead in the Southeast Missouri Lead Mining District could be distinguished by using various analytic techniques. The committee emphasizes that it did not consider exposure pathways, which would need to be considered for identifying which lead sources are contributing to human exposure.

ORGANIZATION OF THIS REPORT

The committee's report is organized into five chapters and three appendixes. Chapter 2 describes possible sources of lead contamination. Chapter 3 describes mechanisms of environmental dispersal of lead, and Chapter 4 proposes investigative strategies for determining sources of lead in various media. Chapter 5 discusses the implications of the committee's strategies specifically for the

BOX 1-2 Statement of Task

An ad hoc committee will examine the extent to which various sources contribute to environmental lead concentrations in areas at or near Superfund sites that are situated within or adjacent to lead-mining areas. The committee will consider whether sources, such as naturally occurring lead, lead-based paint, and the consumer use of products that contain lead, are significant contributors to environmental contamination problems identified through the Superfund site assessment process. In addition, it will consider the extent to which the chemical form of lead in the environmental media can be attributed to particular kinds of sources.

The committee will focus primarily on sources contributing to environmental lead concentrations at or near Superfund sites within or adjacent to the Southeast Missouri Lead Mining District. It also will consider several Superfund sites within or adjacent to other lead mining areas in the United States.

The committee will recommend potential improvements in approaches used for assessing sources of lead contamination at or near Superfund sites, without recommending changes to the clean-up requirements or remediation activities at any designated site. The committee also will identify important research needs.

Southeast Missouri Lead Mining District and describes the research needed for identifying lead sources at Superfund sites associated with mining activities. Appendix A provides information on the challenges of obtaining information on lead sources, Appendix B provides biographic information on the committee, and Appendix C contains the agendas of its open sessions.

REFERENCES

ATSDR (Agency for Toxic Substances and Disease Registry). 1996. Preliminary Public Health Assessment, Big River Mine Tailings Desloge (a/k/a St. Joe Minerals), Desloge, St. Francois County, MO. CERCLIS No. MOD 981126899 [online]. Available: http://health.mo.gov/living/environment/hazsubstancesites/pdf/bigriverminetailings.pdf [accessed June 28, 2017].

ATSDR. 1998. Big River Mine Tailings Superfund Site, Lead Exposure Study, St. Francois County, Missouri [online]. Available: http://health.mo.gov/living/environment/hazsubstancesites/pdf/BRMTLeadExposureStudy98.pdf [accessed June 28, 2017].

Bonne Terre News. 2008. Retrospect on St. Francois County Mining. Pp. 150-151 Chat Dumps of the Missouri Lead Belt, St. Francois County. Adrian, MI: R.E. McHenry.

Buckley, E.R. 1909. Geology of the Disseminated Lead Deposits of St. Francois and Washington Counties, Vol. IX, Part I. Jefferson City, MO: H. Stephens Printing.

EPA (US Environmental Protection Agency). 2017a. Superfund: National Priorities List (NPL) [online]. Available: https://www.epa.gov/superfund/superfund-national-priorities-list-npl [accessed March 10, 2017].

EPA. 2017b. Superfund Cleanup Process [online]. Available: https://www.epa.gov/superfund/superfund-cleanup-process [accessed March 10, 2017].

EPA. 2017c. Search for Superfund Sites Where You Live [online]. Available: https://www.epa.gov/superfund/search-superfund-sites-where-you-live [accessed July 12, 2017].

Guberman, D.E. 2014. Lead. Pp. 90 in Minerals Yearbook: Vol. 1. Metals and Minerals. U.S. Geological Survey [online]. Available: https://minerals.usgs.gov/minerals/pubs/commodity/lead/mcs-2014-lead.pdf [accessed June 28, 2017].

Lee, M. 2006. Symbols of pride. P. XV in Chat Dumps of the Missouri Lead Belt, St. Francois County, 2008. Adrian, MI: R.E. McHenry.

Meyers, H. 1989. Where Did All the Tailings Go? Pp. XI-XII in Chat Dumps of the Missouri Lead Belt, St. Francois County, 2008. Adrian, MI: R.E. McHenry.

MO DNR (Missouri Department of Natural Resources). 2017. Missouri Lead Mining History by County. Missouri Department of Natural Resources [online]. Available: https://dnr.mo.gov/env/hwp/sfund/lead-mo-history-more.htm [accessed February 15, 2017].

Mugel, D.N. 2017. Geology and Mining History of the Southeast Missouri Barite District and the Valles Mines, Washington, Jefferson, and St. Francois Counties, Missouri. Scientific Investigations Report 2016-5173. Reston, VA: U.S. Geological Survey [online]. Available: https://pubs.usgs.gov/sir/2016/5173/sir20165173.pdf [accessed June 28, 2017].

Seeger, C.M. 2008. History of mining in the Southeast Missouri Lead District and description of mine processes, regulatory controls, environmental effects, and mine facilities in the Viburnum Trend Subdistrict. Pp. 1-34 in Hydrologic Investigations Concerning Lead Mining Issues in Southeastern Missouri, M.J. Kleeschulte, ed. Scientific Investigations Report 2008-5140. Reston, VA: U.S. Geological Survey [online]. Available: https://pubs.usgs.gov/sir/2008/5140/pdf/SIR2008-5140.pdf [accessed June 28, 2017].

2

Sources of Lead Contamination

As noted in Chapter 1, environmental lead can have many sources. It can occur naturally or result from mining or other industrial activities or from the production and use of consumer products that contain lead. This chapter describes the various sources that potentially contribute to lead contamination at or near a Superfund site (see Figure 2-1), and Chapter 3 describes the mechanisms by which environmental dispersal might occur.

NATURAL AND MINING-ASSOCIATED MATERIALS

Lead is a widespread, naturally occurring element found throughout the Earth's crust and is commonly enriched in many materials in mining districts. It occurs in virtually all rocks, soils, sediments, and waters at low concentration; however, it can occur at high concentrations over large areas because of natural and human-caused processes. A variety of geologic processes can cause natural lead enrichment, commonly in association with other metal enrichment; in some cases, concentrations of economic interest (ore deposits) are formed. Natural processes that operate at or near the Earth's surface can concentrate lead further or disperse it into the environment. Such processes could include volcanic emissions, forest fires, and various erosional processes. Similarly, human activity, particularly mining of metallic ores, leads to concentration and dispersion (see Figure 2-1).

In most rocks, lead occurs at low concentrations and behaves geochemically like the alkalis and alkaline earths. As such, it is dispersed in common rock-forming minerals, such as feldspars, amphiboles, micas, and sulfates, where it substitutes for potassium, barium, and similar elements (Sahl et al. 1978). At high concentrations, it forms discrete lead-rich minerals, notably sulfides, carbonates, sulfates, and oxysalts (Sahl et al. 1978). Of the roughly 400 lead-rich minerals known (Back et al. 2008), the most common and essentially the only ore minerals are lead sulfide (galena, PbS), lead carbonate (cerussite, $PbCO_3$), and lead sulfate (anglesite, $PbSO_4$). In this section, the committee briefly discusses the types of lead-bearing mineral deposits, the natural distribution of lead, and mining processes that can lead to dispersal of lead in the environment.

Lead-Bearing Mineral Deposits

Many mineral deposits—rocks that contain materials of potential economic interest—and their surrounding materials contain lead in such concentrations as 0.01 to over 10 wt %, which are well above the typical concentrations (5–80 ppm, and usually less than 20 ppm) in rocks and soils found outside mineralized regions (Faure 1998; Smith and Huyck 1999; Rudnick and Gao 2014). As noted above, human activities have led to environmental dispersal of lead; lead concentrations in post-industrial soils and sediments are about 50–100 ppm (Callender 2005), and concentrations at contaminated sites vary widely—up to thousands of parts per million (Gutierrez et al. 2016).

The decision to recover any particular commodity depends on economics, geology, available technology (extraction methods), and other incentives, not simply on the characteristics of the ores themselves. Thus, anywhere from none to most of the lead might be recovered from mined materials. For example, in Southeast Missouri, lead was the primary product from numerous small operations in the Washington County barite district before the 1920s, at which time production of barite became the focus of mining, and lead from that area became a co-product or a byproduct, or was discarded (Wharton 1972; Mugel 2017). Given the various factors that determine whether a commodity is mined, four categories of mineral deposits might be encountered at a Superfund site, as described below.

- Primary lead deposits in which lead is the primary or sole commodity (primary product) that motivates mining. The classic example is the Southeast Missouri Lead Mining District, where the principal value in most deposits is lead.

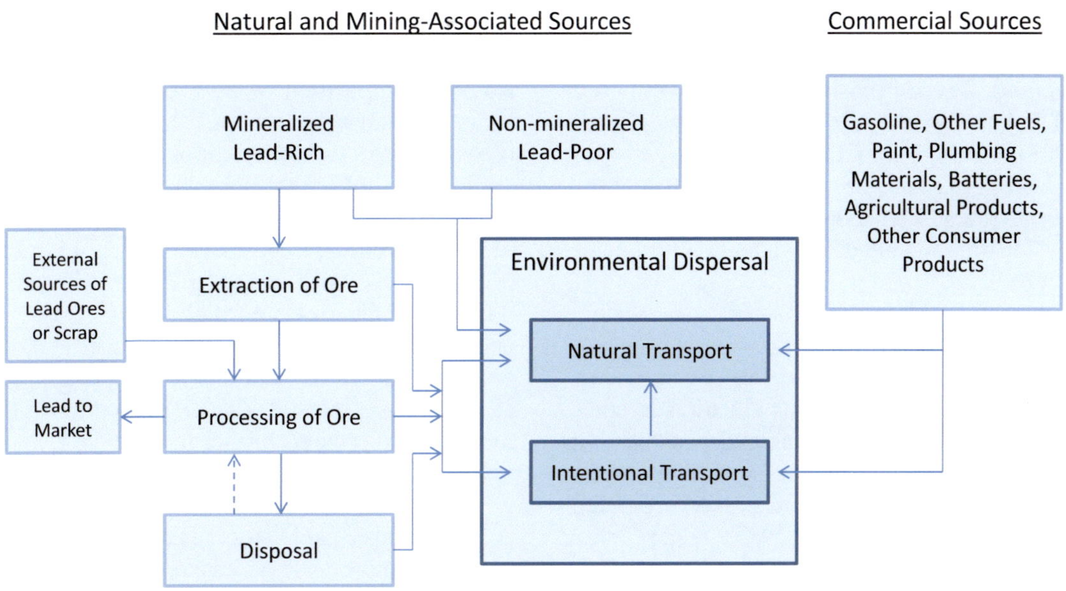

FIGURE 2-1 Natural, mining-associated, and commercial sources can cause lead contamination in the environment through natural or intentional transport. *Natural transport* refers to environmental processes—such as wind, water flow, and soil erosion—that disperse materials; *intentional transport* refers to all human activities that move materials and have the potential for dispersal in the environment.

- Mineral deposits in which lead is a co-product, and its recovery is an essential part of the economics, but production of other metals is also essential. There are a number of examples, including the midcontinent Tri-State district (Missouri-Kansas-Oklahoma), where lead was co-produced with zinc; Coeur d'Alene, Idaho, where most deposits produced lead and silver; and Leadville, Colorado, where lead was co-produced with silver, gold, and zinc.
- Mineral deposits in which lead is recovered as a byproduct because it is economically profitable to do so but not necessary for profitability. Examples include the vein deposits of the Silverton district, Colorado, which include the Gold King, Sunnyside, and related mines that were produced primarily for gold and silver, but where other metals, including lead, were common byproducts.
- Mineral deposits where lead at lower concentrations (commonly less than 0.5%) is mined with the ore but is not recovered. An example is the Ducktown district, Tennessee, which produced copper and iron sulfide (for sulfuric acid).

Metallic mineral deposits form when geologic processes concentrate metals into small volumes of rock from much larger, relatively dilute sources. As in the case of lead, concentrations in ores are typically a thousand times as great as in the source rocks. Lead-bearing deposits form in many geologic settings with a corresponding diversity in ore types and characteristics. The geologic occurrence and the geochemical nature of lead-bearing ores have fundamental effects on how lead gets into the broader environment, whether through natural dispersion or by mining and related human activities (du Bray 1995; Plumlee 1999). The most important lead-bearing deposits and some of their principal characteristics are summarized below.

- *Deposits associated with sedimentary basins.* These deposits constitute the most important lead ores globally and include the Mississippi Valley Type (MVT) ores of the middle of North America (Leach et al. 2005, 2010) and the ores of the Southeast Missouri Lead Mining District. The deposits are typically stratabound (confined to favorable layers), and the rocks that host the deposits are commonly rich in carbonate minerals. The low iron sulfide and high carbonate concentrations of these deposits limit acid production during weathering and thus limit lead solubility and mobility in the surface environment.
- *Deposits associated with hot waters generated from or circulated by magmatism.* These deposits include carbonate replacement, polymetallic veins, and seafloor massive sulfides. They are the most common types of mineral deposits and typically have small to moderate quantities of lead with other base and precious metals (Ridley 2013). They commonly have abundant sulfide minerals, and their host rocks might lack appreciable acid-neutraliz-

ing capacity. Thus, lead can be appreciably mobile under these circumstances.

- *Deposits associated with waters circulated during metamorphism*. These deposits are mainly vein gold deposits with relatively low lead concentrations (Goldfarb et al. 2005), although the Coeur d'Alene silver–lead deposits of Idaho fit this category (Ramos and Rosenberg 2012). The moderate sulfide and carbonate mineral concentrations result in moderate acid production and acid-neutralizing capacity.
- *Residual deposits associated with weathering*. Chemical weathering processes can substantially increase the concentration of lead minerals because of removal of other, more soluble components. This situation is common in the weathered zone of many metal deposits and in the clay-rich soils formed on lead-bearing carbonate host rocks, such as in the Washington County barite district, Missouri (Wharton 1972; Kaiser et al. 1987). Acid production is minimal, and neutralization capacity is moderate.

Geologic differences are important because they lead to different environmental consequences, they allow possible fingerprinting of sources, and they influence the choice of mining and processing methods by which lead and other commodities are produced. Among the important ore characteristics are the capacity to produce acid waters when weathered, the content of toxic elements, and the geologic form and hydrologic setting of the deposits (du Bray 1995; Plumlee 1999; Smith and Huyck 1999; Smith 2007). As summarized above, acid production correlates with the abundance of iron sulfide minerals (pyrite, pyrrhotite, and marcasite), whereas acid-neutralizing capacity is highest in carbonate-hosted deposits, such as those of the midcontinent and many igneous-related replacement deposits. Conversely, seafloor massive sulfide deposits and many vein deposits have relatively high sulfide concentrations and host rocks that are poor acid neutralizers, and these characteristics lead to the wide availability of acid and dissolved metals in groundwater and surface waters.

Lead concentrations (and other element concentrations) vary greatly among ore deposit types and within individual deposits or districts (du Bray 1995; Plumlee 1999; Smith and Huyck 1999). Some ore deposits and ore districts have relatively uniform compositions and host rocks (for example, many MVT deposits and districts), but many exhibit strong spatial variation (zoning) in metals, other elements, and the nature of the local host rocks (for example, magmatic-related hydrothermal systems; Ridley 2013). Lead concentrations can span several orders of magnitude, depending on position; for example, ores at Bingham, Utah, ranged from greater than 10% in the carbonate rock-hosted part of the system to less than 0.1% in the igneous rock-hosted copper-rich core. Because the geochemical signatures vary with deposit type and location of a particular ore-forming system, they can be used to aid in the forensic identification of sources.

A particularly useful geochemical variable is the isotopic composition of lead itself (Box 2-1). Extensive study of lead isotopes in ores shows that different deposit types and regions have compositions that reflect the geologic history of the terrain and the processes involved in ore formation (Doe and Zartman 1979). For example, ores of the midcontinent have distinctively high ratios of the radiogenic isotopes of lead (^{206}Pb, ^{207}Pb, and ^{208}Pb) to non-radiogenic, or primordial, lead (^{204}Pb).[1] Such differences have the potential to distinguish geologic sources and even individual districts and deposits (Goldhaber et al. 1995; Sangster et al. 2000). Application of isotopic differences is discussed further in Chapter 4, and case studies are provided in Chapter 5.

Over 200 mining districts in the United States have produced appreciable amounts of lead (Long et al. 1998), and many more have accessory lead minerals. However, the principal lead production has come from only a large handful of districts. The most important are the Mississippi Valley ores of Southeast Missouri and the Missouri–Kansas–Oklahoma tristate district; the vein silver–lead ores of the Coeur d'Alene, Idaho; and various igneous-related polymetallic carbonate replacement deposits of Colorado (Leadville) and Utah (Bingham, Park City, and Tintic). Of those districts, Southeast Missouri (Box 2-2; Figure 2-2) has produced more than the others combined (Long et al. 1998) and collectively contains the world's largest known concentration of lead (Appold and Garven 1999).

Natural Distribution of Lead

When lead deposits form from hydrothermal fluids, the lead concentration rarely terminates abruptly at the edge of an ore body but rather falls off gradually; this is primary dispersion.[2] The original geologic dispersion reflects the facts that ore-forming fluids typically invade a considerable volume of rock and that the metal-trapping (precipitating) process is commonly inefficient (Rose et al. 1979). Such primary distributions can be highly focused, around a few meters or less, if the flow of mineralizing fluids was strongly focused (Barnes and Lavery 1977). In other cases, where fluid flow and conditions for precipitation were not localized, tremendous volumes of rock over large areas (tens to thousands of square kilometers) might contain anomalous metals. For example, minor quantities of galena and sphalerite (ZnS) are widely distributed in the

[1]Radiogenic indicates an isotope that has been created from radioactive decay.

[2]The edge of an ore body is an economic boundary that is influenced by a combination of geologic and technologic factors. In many cases, what might be waste under one set of conditions can become ore under more favorable conditions, or vice versa.

BOX 2-1 Lead Isotopes: A Tool for Source Attribution

Lead-isotope ratios in geologic materials vary because lead isotopes are produced by the radioactive decay of uranium (U) and thorium (Th) over geologic timescales. ^{235}U decays to ^{207}Pb, ^{238}U decays to ^{206}Pb, and ^{232}Th decays to ^{208}Pb, whereas ^{204}Pb is primordial and has no radiogenic parent. Because of contrasting geologic histories, ages, and U/Pb and Th/Pb ratios, there is a wide range of lead-isotope ratios among lead-bearing ore deposits and other geologic materials (see, for example, Sangster et al. 2000; Faure and Mensing 2004). Lead isotopes are not altered in mining, processing, or manufacturing; thus, ratios of isotopes from a given source are the same whether in natural materials, lead-mining waste, or commercial products (provided that a single source has been used in the manufacturing process). A simple isotope mixing equation describes the lead isotopic composition of lead from mixtures of sources (Faure and Mensing 2004).

Lead-isotope ratios are often reported as ^{206}Pb/^{204}Pb, ^{207}Pb/^{204}Pb, and ^{208}Pb/^{204}Pb, and plots of these ratios give the most complete differentiation between ore deposits and between sources of environmental lead. The full complement of lead-isotope abundances is measured with thermal ionization mass spectrometry and multi-collector inductively coupled plasma mass spectrometry (ICP-MS). However, some studies that use single-collector ICP-MS do not measure the lower-abundance ^{204}Pb (which makes up about 1.5% of total lead) with adequate precision; therefore, to use all available lead-isotope data in the literature, environmental applications of the ratios often use the ^{206}Pb/^{207}Pb and ^{208}Pb/^{207}Pb ratios or simply the ^{206}Pb/^{207}Pb ratio alone for comparisons between lead sources. Sangster et al. (2000) used a statistical analysis to demonstrate that ^{204}Pb was not critical for discriminating between most lead sources.

Numerous studies have used lead-isotope ratios to distinguish between potential sources of lead in soils and atmospheric and household dust (see, for example, Gulson et al. 1981; Graney et al. 1995). A smaller number of studies have included lead isotopic analyses of human blood or urine; that approach has in some cases allowed elimination of some potential sources and in other cases has allowed identification of the sources of lead exposure (see, for example, Manton 1985; Gwiazda and Smith 2000).

BOX 2-2 Geology of Southeast Missouri Lead Mining District

In Southeast Missouri, tremendous quantities of lead are present in flat-lying Cambrian and Ordovician sedimentary rocks (about 450–550 million years old) that surround and lap onto the edges of an ancient island represented by the 1.5-billion-year-old igneous rocks of the St. Francois Mountains (Figure 2-2). Lead and other components—such as barium, zinc, copper, cobalt, and nickel—were added by warm (about 100°C) saline fluids derived from adjacent sedimentary basins to the south. The largest lead deposits occur in the oldest sedimentary rocks, mainly the Bonneterre Dolomite and underlying Lamotte Sandstone near where they terminated against the older igneous rocks. The yellow-clay soils that developed on the dolomite were enriched in lead; they were referred to as the "bonne terre" by early miners, reported to contain abundant galena in the upper few tens of meters, and attracted preindustrial development (Winslow 1894; McHenry 2006, p. 362). After the 1860s, the Bonneterre ores were hard-rock ores that were mined underground.

In contrast with the majority of the Bonneterre ores, the lead-bearing barite mineralization in the younger rocks of the Potosi Dolomite and overlying formations consists of residual deposits that formed from widespread but low-grade primary sulfide mineralization, which was concentrated by chemical weathering in residual soils. Because of lead's relatively low solubility, galena and its oxidation products (cerussite and anglesite) remained in the clay-rich soils with similarly insoluble barite (Wharton 1972; Kaiser et al. 1987). Both commodities were readily recovered from numerous shallow mines, many of them open pits across a wide region, which date back into the early 19th century. Early mining and local mining focused on lead, but from the 1920s to the 1980s these residual deposits were mined by industrial methods for barite (MO DNR 2017; Mugel 2017).

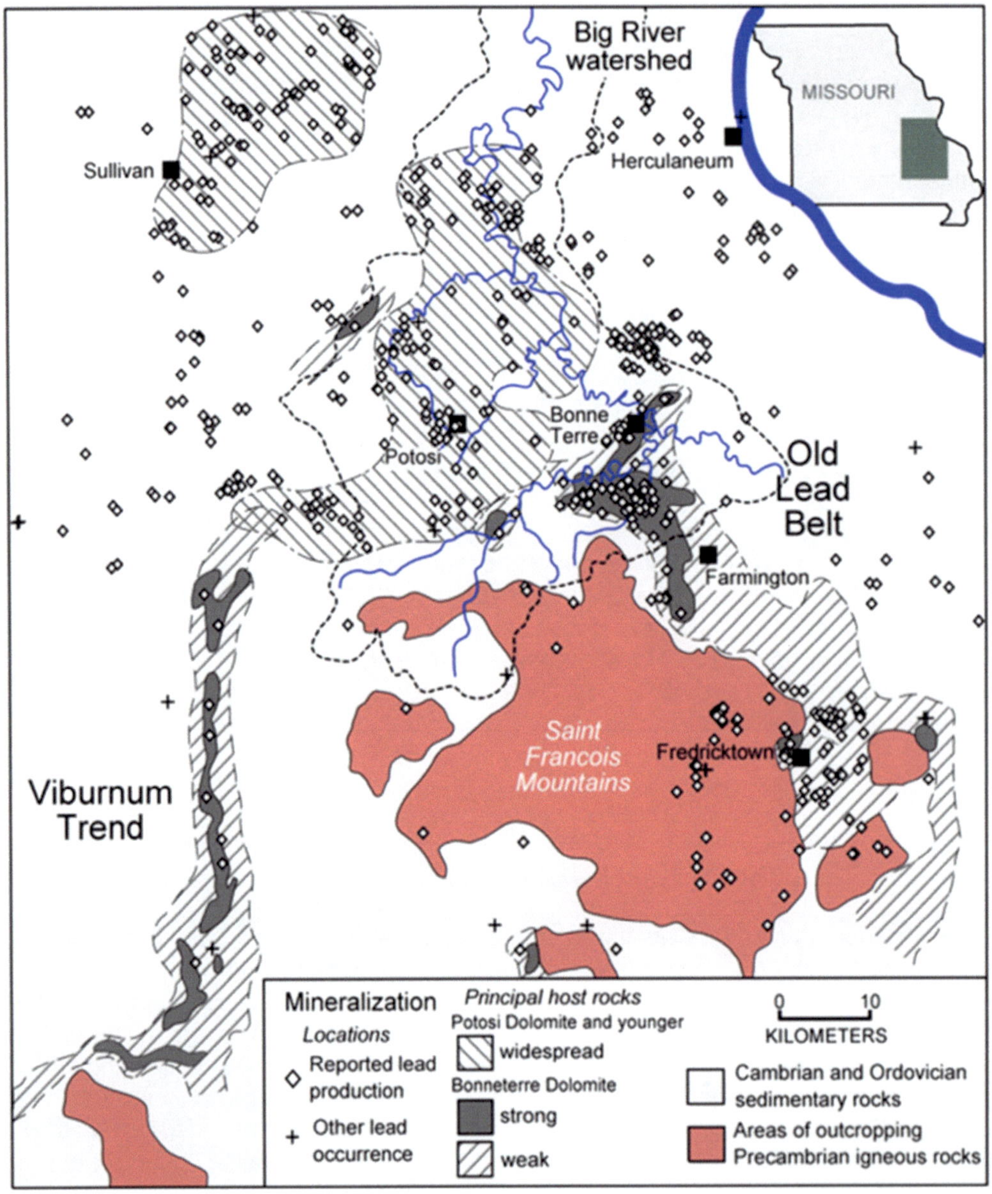

FIGURE 2-2 Distribution of lead mineralization in Southeast Missouri and related features, including the Big River watershed. The thin blue lines are rivers, and the thick blue line is the Missouri–Illinois border. Lead mineral occurrences are from the US Geological Survey Mineral Resources Data System (USGS 2015). High-grade lead mineralization occurs principally in the Bonneterre Dolomite and underlying Lamotte Sandstone, but low-grade lead mineralization is widespread with barite in the residual ores of the overlying Potosi Dolomite and younger formations. Lead mineralization is widely present at the surface in the eastern and northern parts of the region but is essentially restricted to the subsurface along the Viburnum Trend to the west. The boundaries for the Bonneterre-Dolomite mineralization were redrawn from a review by Zientek and Orris (2005), which did not define *strong* and *weak* in the figure inset. However, *strong* appears to refer to regions of mineralization that incorporate areas where there has been substantial lead mining, and *weak* to regions that rarely had any lead production but commonly had lead mineralization in trace quantities.

Paleozoic sedimentary rocks of the midcontinent, where they represent the sparse products of the metalliferous fluids that flowed regionally along these sedimentary layers (McLimans et al. 1980; Goldhaber et al. 1995; Leach et al. 2005; Figure 2-2).

Weathering and erosion of primary metal concentrations lead to dispersion at or near the Earth's surface. Given the relatively low solubility of lead in most surface fluids, chemical weathering concentrates lead in residual materials in gossans (weathered ores) and clay-rich soils (Rose et al. 1979; Smith and Huyck 1999). The residual materials are ultimately eroded (mechanically removed by running water) and contribute to the overall sediment load in rivers and streams. Only in sulfide-rich deposits that have poor neutralizing capacity is lead dispersed mainly in solution. Environmental dispersal of lead through those various processes is discussed in detail in Chapter 3.

Concentration and Dispersal of Lead by Mining

Lead is concentrated to yield a product but also dispersed by mining (Figure 2-1). Mining begins by physically extracting ore; this is followed by processing of the ore to obtain a concentrate of lead minerals and then by smelting to obtain metallic lead, which is ultimately shipped to market. At each stage, materials are discarded. In the extraction process, waste consists of materials that are moved to mine the ore but that lack economic concentrations of metal. Waste rock can vary greatly in mineralogy, composition, and size. Because the distinction between ore and waste is economic, large volumes of waste can have elevated lead concentrations compared with those of ordinary rocks. Indeed, what is waste under one set of conditions or at one time might be ore in other circumstances, and reprocessing for environmental or economic reasons is common in many long-lived mining districts. Historically, waste disposal was driven by what was easiest; thus, waste was generally discarded in the most efficient manner for the time and place, such as on the surface in dumps or replaced in mine openings. However, by the latter part of the 20th century, waste rock disposal was subject to regulatory controls and handled systematically in a mine's plan of operation.

Processing of lead-bearing ores typically involves physical separation as is the case with all lead minerals. However, some lead-bearing ores might be chemically processed; for example, a cyanidation process is used to treat lead-bearing gold deposits. Tailings are the residua of mineral processing after much of the valuable content has been removed in the form of a concentrate. The efficiency of concentration reflects available technology and economics, and recovery of valuable components can vary from less than 50% to 95%. For example, early and primitive techniques used hand-picking or simple gravitational separation with rockers to upgrade the content of the valuable fraction of the ore; the techniques were inefficient and often resulted in less than 50% recovery, and the balance went into the mine waste or tailings (Mugal 2017). Given the high lead content of ores, tailings typically contain lead concentrations of a few hundred to thousands parts per million, depending on recovery efficiency.

Most mineral processing depends on the characteristic properties of ore minerals that allow separation of the valuable constituents from the unwanted minerals, for example, by using density or distinctive surface properties. Gravity-based methods operate by separating dense ore minerals—such as galena, cerussite, and barite—from a less-dense waste fraction. If the lead-bearing materials are not consolidated, as is the case in residual ores, simple washing and gravity separation might suffice. For example, such methods were widely used in the barite-lead districts of Southeast Missouri, notably in Washington County. In that case, the efficiency of recovery depended on the main product—barite or lead—and, in many cases, a substantial fraction of lead was not recovered and thus discarded in the tailings. See Box 2-3 for additional information on the history of lead mining in Southeast Missouri (Wharton 1972; Mugal 2017).

In hard-rock mining, ores require crushing (milling) to liberate the valuable minerals from the unwanted gangue (waste) minerals. The degree of crushing depends on the nature of the ore, particularly on the grain size of the ore minerals. Except in unusually coarse-grain ores, efficient separation (for example, greater than 90%) commonly requires crushing to fine sand or smaller. How finely to crush the ore is a tradeoff in cost vs recovery. After the ore is milled, the ore minerals are separated from the waste to make an ore concentrate. Gravity separation is still commonly applied, but most sulfides are recovered through selective froth flotation. In froth flotation, hydrophobic surfactants specific to particular minerals are added to a slurry of the crushed ore. When air is bubbled through the slurry, the surface-coated minerals attach to the air bubbles and float to the surface. The metal-laden froth is continuously removed and dried to leave a high-grade mineral concentrate. The concentrate is the feedstock for the final steps in metal production.

Tailing characteristics—such as mineralogy, grain size, and chemical composition—reflect differences in processing, economics, and the broad social and regulatory environment. Over time, tailings disposal has seen an increasing focus on engineered impoundments. In recent decades, designed and built impoundments control not only the bulk tailings themselves but their potential water and dust effects. Early mining operations paid little or no attention to disposal methods; often, there was no effort at impoundment, and some tailings were discarded directly into streams. With time, tailings were typically discarded in designated areas, but often the areas were not designed to control slope failure, other types of erosion, wind effects, and water discharges.

The final steps in the recovery of lead are broadly termed smelting; they reflect conversion of the ore concentrate into metal and various waste products, including slag, gases, and dust (Vignes 2013). Smelter feed can include both primary sources (concentrates) and secondary sources (from recycling). As smelters became larger industrial operations through the 20th century, their feedstock evolved from local sources (one or more nearby mines) to a mix of sources.

Depending on the nature of the feed (recycled metals or ore concentrate), the smelting process begins with pretreatment that might include mechanical separation (secondary sources) or drying, calcining, or roasting (primary sources). The materials are then smelted and converted to metal by reduction from the sulfide or oxide form. The molten waste product (slag) contains iron, silica, calcium, some residual metal, and other minor components. Sulfide smelting involves passage of gases that carry off the sulfur, mainly as sulfur dioxide. The emitted gases also contain appreciable amounts of particulate matter in the form of melts or solids. As with mine waste and tailings, early smelting operations were typically inefficient and operated with no emission controls. Gases, slag, and dust were discharged directly into the environment. In more recent times, particularly with the passage of the Clean Air Act and resulting regulation, smelter emissions have been greatly reduced by treatment of sulfur-bearing gases and by control of dust and other particles. Slag is typically disposed of in heaps, where it is poured, cools, and solidifies. Ultimately, the metals produced by smelting might be further processed in a refinery to recover precious metals (notably silver in the case of lead) or to remove impurities that are not wanted in the final salable product. Like the earlier stages, solid and liquid refinery wastes have had a checkered history of disposal and might still contain appreciable metals. After smelting and mining, lead is distributed to other regions where it is converted into manufactured or chemical products, providing other opportunities for lead dispersion in areas that are typically well outside mining areas.

OTHER SOURCES OF ENVIRONMENTAL LEAD

There are possible sources of environmental lead in addition to natural sources and those associated with mining (Figure 2-1). Indeed, the 2014 National Emissions Inventory estimated that less than 1% of atmospheric lead emissions in the United States was the result of mining activities (Table 2-1). Mining-related emissions of lead in Missouri accounted for one-third of the national mining-related emissions but only 10% of the total lead emissions in the state. This section briefly discusses other sources of lead that could contribute to environmental contamination.

BOX 2-3 History of Lead Production in Southeast Missouri

The history of lead production in Southeast Missouri illustrates the challenges in attributing sources of lead to mining compared with other sources (MO DNR 2017; Mugel 2017). Early mining and smelting techniques were primitive; surface disturbance was extensive, and there were many local smelting operations. Lead minerals were recovered by hand, including primitive gravity separation techniques, and losses ranged from as little as 10% to greater than 50% of the total contained lead, which then ended up in mine waste and primitive tailings. Cerussite and galena were smelted locally with a series of open-air techniques that also failed to recover tens of percent of the lead contained in the ore mineral concentrates.

As noted in Chapter 1, industrial-scale mining began in the 1860s with development of the major underground ore bodies in the Old Lead Belt (Figure 2-2), but considerable small-scale surface mining continued by local landowners through the 1930s. Conversion to barite mining in the 1920s in the Washington County barite district was accompanied by a shift from largely local operations for lead to industrial-scale production of barite, which led the world in barite production for decades (Wharton 1972). High-grade ores in the Old Lead Belt were gradually exhausted with the last mine closing in 1972. By that time, the Viburnum Trend, or New Lead Belt (Figure 2-2), was in full production. Those ores, which are completely covered by younger rocks, were discovered in 1955, and the first production was in the middle 1960s. In contrast with the other areas in Southeast Missouri—where there was a century or more history of mining, processing, and smelting—the industrial-scale operations of the Viburnum Trend were developed in a far more consolidated manner with designed mine-waste dumps and tailings impoundments. In parallel with mining and milling, smelting operations were progressively more centralized, and smaller operations closed. The last primary smelter, at Herculaneum, Missouri (Figure 2-2), closed in 2013, although the Doe Run Company continues to run one of the world's largest secondary lead smelters for recycling of lead-acid batteries and other scrap at Boss, Missouri.

TABLE 2-1 US National Lead Air Emission Estimates for 2014[a]

Sector or Source	US Lead Emissions (tons/yr)	Sector or Source Fraction (%)
Energy, all sources	88.9	12.2
Industrial processes		
Mining[b]	4.9	0.7
Metals[c]	137.8	18.9
All other	20.2	2.8
Transportation		
Aviation	456.3	62.5
All other	5.2	0.7
All other sources or sectors	17.0	2.3
Total	730.2	100.0

[a]Data compiled from the 2014 National Emission Inventory.
[b]*Mining* refers to mineral-extraction activities.
[c]*Metals* refers to smelting, refining, and other metal-processing activities.
Source: EPA (2017).

Gasoline and Other Fuels

In the United States, lead has been added to gasoline and other fuels since 1929 to prevent engine knocking. Mielke (1999) estimated the total consumption of lead in gasoline in the United States from 1929 to 1986 as 5.9×10^6 metric tons. The use of alkyl–lead additives in automobile fuels has been nearly eliminated since US Environmental Protection Agency (EPA) promulgated guidelines in the 1970s that mandated that new cars run on unleaded gasoline. The sale of leaded automobile fuels was discontinued in 1986, although a small amount of alkyl–lead continues to be used in racing sports and by owners of classic cars. The sharp decline in lead consumption in gasoline (Figure 2-3) resulted in a 98% decrease in total national lead emissions from 1970 to 1995 (EPA 2013).

Alkyl–lead additives are still used in aviation fuel, specifically for small piston-engine aircraft. Although emissions that result from the operation of those airplanes was a tiny fraction of total atmospheric lead emissions in most of the 20th century, they accounted for more than half the national lead emissions by 2008 (Rao et al. 2013) and for nearly two-thirds in 2014 (Table 2-1). Aircraft-related lead emissions are highly variable; areas near airports experience the highest loadings.

Although automobile-related lead emissions are now near zero, the legacy lead that remains in soils and sediments is a potentially important lead source at or near Superfund sites. Smith (1976) found that lead concentrations in the top 5 cm of roadside soils were up to 30 times the concentrations in soils away from the roadside. Emissions of gasoline lead, however, were widely dispersed around the country and resulted in deposition and accumulation of lead in soils and surface waters even in remote areas. Johnson et al. (1995) estimated that the total lead deposition in precipitation at the Hubbard Brook Experimental Forest in New Hampshire in 1926–1989 was 8.66 kg/ha. Because Hubbard Brook is remote from point sources of lead pollution, the lead deposition was interpreted as an estimate of gasoline-lead input into the forest. Lead is readily adsorbed in soil organic matter and clay particles (see Chapter 3), so nearly all the lead remains in the soil and makes up a large fraction (about 10%) of total lead in the soil (Johnson et al. 2004). At Hubbard Brook, the mass of the O soil horizon (surface soil layer), where most of the atmospherically deposited lead is stored, is 8.7 kg/m^2 (Johnson et al. 1991); thus, gasoline lead might account for up to 100 ppm of the total lead in the O soil horizon. Furthermore, the 1926–1989 atmospheric lead input was similar in magnitude to that of total lead released from soil minerals by chemical weathering in the 12,000–14,000 years since the last glacial retreat. The burden of gasoline-derived lead in soil can thus be quantitatively important in studies aimed at lead-source apportionment.

Paint

The use of lead in paints in the United States dates back to the colonial era. White lead provided a consistent base color for interior and exterior paints and made the paint more durable and easy to wash (Browne 1930; Hallett and Rose 1935). Widespread use of lead in US paint manufacturing began in 1884 and peaked in the 1920s,

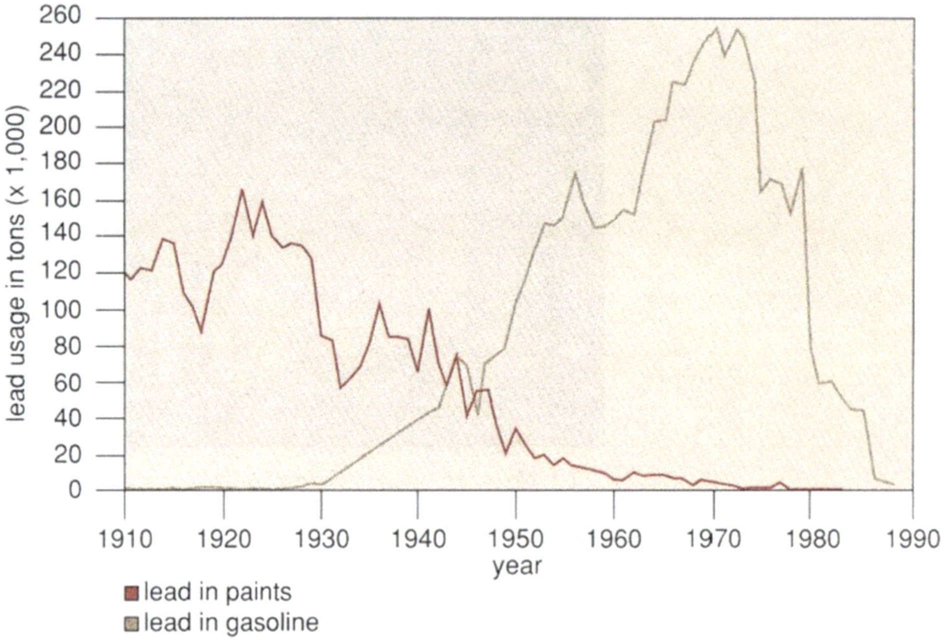

FIGURE 2-3 Annual consumption of lead in the United States in paint and automobile gasoline. Source: Mielke (1999). Reprinted with permission; copyright 1999, *American Scientist*.

but lead-based paints continued to be widely used, especially on exterior surfaces, into the 1970s (Figure 2-3). Baltimore, Maryland, passed the first municipal law that banned the use of interior lead paints in 1951. In 1978, the federal government banned the sale of all lead-based paints. The total consumption of lead by the US paint industry was similar in magnitude to that used in gasoline additives for automobile fuels (Mielke 1999). Therefore, as with gasoline, old paint could represent a quantitatively important source of lead in some soils.

Because of the durability of lead paints and the tendency to paint over, rather than remove, previous paint layers, a large amount of lead is associated with older buildings and with construction debris. Chipping, physical abrasion, and breakdown of paint surfaces by atmospheric chemicals, such as ozone and nitrogen dioxide (Edwards et al. 2009), lead to the release of particulate, aerosol, and gaseous lead from interior and exterior surfaces of residential, commercial, and industrial structures. Soils near buildings constructed before the 1970s often have high lead concentrations, especially near the drip lines where water runs off rooftops. In an extensive study in New Orleans, lead concentrations were substantially higher in soils collected along drip lines of residential homes than in soils near streets or in parks (Mielke 1994; Mielke and Reagan 1998).

Paints used for road markings have also historically contained lead. Abrasion of those paints by vehicles results in road dust with high lead concentrations, much of which is deposited along the sides of the road (Murakami et al. 2007). Some oil paints used by artists continue to contain lead.

Plumbing Materials

Lead pipes were used extensively in indoor plumbing and for residential service lines in most American communities well into the 20th century. Most municipal building codes now prohibit new installations of lead piping, but many homes and older buildings still have lead risers inside and lead service lines outside. In some municipalities—notably Flint, Michigan—lead service lines deliver large amounts of the domestic water supply. Furthermore, lead solder has been used to join copper pipes and pipe fittings and might also be a source of lead in domestic plumbing systems. Corrosion and chipping of the pipes are potential sources of lead contamination in drinking water and in soils outside residential and commercial buildings.

Demolition of aging buildings can result in large releases of lead into the environment if lead pipes and other building materials that contain lead are not removed. Construction debris is thus an important source of lead in many reclaimed urban sites. Superfund sites associated with industrial activities dating back to the first half of the 20th century or earlier might thus have substantial stocks of lead from plumbing and building materials.

Batteries

The manufacture of batteries is one of the largest industrial uses of lead globally. The lead-acid ($Pb–PbO_2$) battery remains the standard means for generating electricity in most vehicles. In the United States, there is an efficient recycling program for lead-acid batteries of all types, which has greatly reduced the amount of battery lead released into the environment generally and deposited in municipal landfills specifically. Smith (2004) estimated a recycling efficiency of 95% for lead in batteries in the United States. The domestic recycling efficiency was estimated to be 77% owing to the export of spent batteries (primarily to Mexico) for recycling. Papp (2013) used a different accounting approach and reported a similar value of 81% for 2011. Wilburn (2014) estimated that 95.5% of the mass of lead-acid batteries was recovered from the municipal waste stream in 2000–2010.

Estimates of lead recycling do not account for long-term storage of batteries before recycling. Substantial lead release can occur from batteries experiencing corrosion. Battery lead might thus be an important source of contamination at sites that host automotive repair businesses.

Agricultural Amendments

Mining wastes have been used as agricultural amendments because metals are commonly associated with carbonate or phosphate rocks (Davies et al. 1993). In the Southeast Missouri Old Lead Belt, for example, chat left over from the extraction of lead from the Bonneterre limestone was crushed and sold for use as agricultural lime through much of the 20th century. Much of that material was used in the surrounding region, but some was shipped by train and barge to towns and ports along the Mississippi River.

Pesticide use has also contributed to environmental contamination with lead. Arsenicals were widely used as pesticides and insecticides in the United States until they were banned in 1988. Lead arsenate—$PbHAsO_4$ or $Pb_5(OH)(AsO_4)_3$ (EPA 1986)—is believed to have been the most commonly used arsenical pesticide (Peryea 1998). It was used for a wide array of crops, but its use in orchards, especially for grapefruit and apples (EPA 1986; Peryea 1998), was particularly widespread. Lead arsenate was typically applied by handgun spraying, and application rates depended on the crop species and the target pest (Peryea 1998).

Extensive use of lead arsenate in orchard lands has necessitated remedial action when those lands have been reclaimed for residential use. Schooley et al. (2008) highlight cases from Virginia, North Carolina, Washington, New Jersey, and Wisconsin in which soil lead concentrations of up to 800 ppm were observed in former apple orchards. Contamination at pesticide manufacturing sites has also been documented. The Industri-plex Superfund site in Woburn, Massachusetts—the site of a former chemical plant that made arsenate pesticides—contained soils contaminated with lead, arsenic, and other metals (DOJ 2012). Ayuso et al. (2004) were able to use lead isotopic analysis to characterize the isotopic signature of lead-arsenate pesticides and concluded that the use of these products could be an important source of lead in soils of formerly agricultural lands in New England.

Other Consumer Products

Lead is an important component of a number of consumer products. Perhaps most notable is ammunition for firearms. Lead shot was nearly ubiquitous until the early 1990s. In 1991, its use in waterfowl hunting was banned. In January 2017, the use of lead shot on federal wildlife refuges was also banned—a regulation overturned in February 2017. The burden of lead in soils and wetlands in or near shooting ranges and popular hunting sites can be high enough to be a threat to wildlife, especially waterfowl and raptors (Kendall et al. 1996; Fisher et al. 2006). Lead sinkers can also be an important source of lead in lakes and rivers that are frequented by fishermen (Jacks et al. 2001; Scheuhammer et al. 2003).

Lead-containing toys, jewelry, and decorative arts are other potential sources that could be important in residential areas. Antique toys and toys that are imported from countries where the use of lead paint is permitted are particularly likely to have high lead concentrations. In some cases, the lead is found in the paint used to decorate the toys. Lead might also be incorporated into the plastics of inexpensive toys. Because lead helps to make plastic more malleable, it is likely to be found in soft plastic toys. Similarly, other household plastics, such as vinyl window blinds, can be an important source of household lead. Sunlight and heat can degrade the material and result in release of dust that has high lead concentrations (EPA 2016). In older homes, lead might also have been used in counterweights in windows.

Lead is abundant in several forms of glass. Leaded glass, which is used for artisan crystal and stained glass, is composed of 12–60% lead oxide, and opal illumination glass contains an average of 3.0% lead oxide (Spinosa et al. 1979). In the late 19th and early 20th centuries, Southeast Missouri hosted an active glass industry, including extraction of local sand for glass-making (Burchard 1907). Many glazes and solders have high lead concentrations. Leaded glazes are used because they provide a smooth finish that highlights bright colors. Terra cotta pottery from Latin America, such as Mexican bean pots, often has high lead concentrations, as do some highly decorated ceramics from Asia. Lead solder is often used in electronic parts and in inexpensive jewelry, and lead is used to make jewelry heavier.

The interested reader is directed to Tables 3-3 and 3-4 in EPA (1998), which contain comprehensive lists of industrial and commercial uses of lead alloys and lead compounds, respectively. Lead derived from the decay of consumer products or present in demolition debris is a potential source to consider in attribution studies in lead-contamination investigations, especially those associated with exposures in the residential environment.

REFERENCES

Appold, M.S., and G. Garven. 1999. The hydrology of ore formation in the southeast Missouri district: Numerical models of topography-driven fluid flow during the Ouachita orogeny. Econ. Geol. 94(6):913-935.

Ayuso, R., N. Foley, G. Robinson Jr., G. Wandless, and J. Dillingham. 2004. Lead Isotopic Compositions of Common Arsenical Pesticides Used in New England. U.S. Geological Survey Open File Report 2004-1342. Reston, VA: U.S. Geological Survey [online]. Available: https://pubs.usgs.gov/of/2004/1342/2004-1342.pdf [accessed June 20, 2017].

Back, M.E., J.A. Mandarino, and M. Fleischer. 2008. Fleischer's Glossary of Mineral Species, 2008. Tuscon, AZ: Mineralogical Record Incorporated.

Barnes, H.I., and N.G. Lavery. 1977. Use of primary dispersion for exploration of Mississippi Valley-type deposits. J. Geochem. Explor. 8(1-2):105-115.

Browne, F.L. 1930. Effect of priming-coat reduction and special primers upon paint service on different woods. Ind. Eng. Chem. 22(8):847-854.

Burchard, E.F. 1907. Glass sand of the Middle Mississippi Basin. Pp. 452-472 in Contributions to Economic Geology, 1905, S.F. Emmons, and E.C. Eckel, eds. U.S. Geologic Survey Bulletin 285N [online]. Available: https://pubs.usgs.gov/bul/0285n/report.pdf [accessed June 28, 2017].

Callender, E. 2005. Heavy metals in the environment—Historical trends. Pp. 67-105 in Environmental Geochemistry, Treatise on Geochemistry Vol. 9, B. Sherwood Lollar, ed. Amsterdam: Elsevier.

Davies, B.E., C.F. Paveley, and B.G. Wixson. 1993. Use of limestone wastes from metal mining as agricultural lime: Potential heavy metal limitations. Soil Use Manag. 9(2):47-52.

Doe, B.R., and R.E. Zartman. 1979. Plumbotectonics, the phanerozoic. Pp. 22-70 in Geochemistry of Hydrothermal Ore Deposits, 2nd Ed., H.L. Barnes, ed. New York: John Wiley.

DOJ (U.S. Department of Justice). 2012. Companies Agree to $4.25 Million Natural Resource Damages Settlement at Industri-Plex Superfund Site, Woburn, Mass. Press Release 12 603. May 10, 2012 [online]. Available: https://www.justice.gov/opa/pr/companies-agree-425-million-natural-resource-damages-settlement-industri-plex-superfund-site [accessed August 3, 2017].

du Bray, E.A., ed. 1995. Preliminary Descriptive Geoenvironmental Models of Mineral Deposits. U.S. Geological Survey Open-File Report 95-831. Reston, VA: U.S. Geological Survey [online]. Available: https://pubs.usgs.gov/of/1995/0831/report.pdf [accessed June 20, 2017].

Edwards, R.D., N.L. Lam, L. Zhang, M.A. Johnson, and M.T. Kleinman. 2009. Nitrogen dioxide and ozone as factors in the availability of lead from lead-based paints. Environ. Sci. Technol. 43(22): 8516-8521.

EPA (U.S. Environmental Protection Agency). 1986. Lead Arsenate. EPA Pesticide Fact Sheet 12/86 [online]. Available: https://quicksilver.epa.gov/work/03/2087981.pdf [accessed June 20, 2017].

EPA. 1998. Locating and Estimating Air Emissions from Sources of Lead and Lead Compounds. EPA-454/R-98-006. U.S. Environmental Protection Agency, Office of Air Quality Planning and Standards, Research Triangle Park, NC. May 1998 [online]. Available: https://www3.epa.gov/ttnchie1/le/lead.pdf [accessed June 20, 2017].

EPA. 2013. Integrated Science Assessment for Lead. EPA/600/R-10/075F. U.S. Environmental Protection Agency, National Center for Environmental Assessment, Research Triangle Park, NC.

EPA. 2016. Lead. Home Danger Zone Finder [online]. Available: https://www.epa.gov/lead/home-danger-zone-finder-0 [accessed August 4, 2017].

EPA. 2017. 2014 National Emissions Inventory (NEI) Data [online]. Available: https://www.epa.gov/air-emissions-inventories/2014-national-emissions-inventory-nei-data [accessed June 20, 2017].

Faure, G. 1998. Principles and Applications of Geochemistry: A Comprehensive Textbook for Geology Students, 2nd Ed. Upper Saddle River, NJ: Prentice Hall.

Faure, G., and T.M. Mensing. 2004. Isotopes: Principles and Applications, 3rd Ed. Hoboken, NJ: Wiley. 493 pp.

Fisher, I.J., D.J. Pain, and V.G. Thomas. 2006. A review of lead poisoning from ammunition sources in terrestrial birds. Biol. Conserv. 131(3):421-432.

Goldfarb, R.J., T. Baker, B. Dubé, D.I. Groves, C.J.R. Hart, and P. Gosselin. 2005. Distribution, character, and genesis of gold deposits in metamorphic terranes. Pp. 407-450 in Economic Geology: One Hundredth Anniversary Volume 1905-2005, J.W. Hedenquist, J.F.H. Thompson, R.J. Goldfarb, and J.P. Richards, eds. Littleton, CO: Society of Economic Geologists.

Goldhaber, M.B., S.E. Church, B.R. Doe, J.N. Aleinikoff, J.C. Brannon, F.A. Podosek, E.L. Mosier, C.D. Taylor, and C.A. Gent. 1995. Lead and sulfur isotope investigation of Paleozoic sedimentary rocks from the southern Midcontinent of the United States; Implications for paleohydrology and ore genesis of the Southeast Missouri lead belts. Econ. Geol. 90(7):1875-1910.

Graney, J.R., A.N. Halliday, G.J. Keeler, J.O. Nriagu, J.A. Robbins, and S.A. Norton. 1995. Isotopic record of lead pollution in lake sediments from the northeastern United States. Geochem. Cosmochim. Acta 59(9):1715-1728.

Gulson, B.L., K.G. Tiller, K.J. Mizon, and R.H. Merry. 1981. Use of lead isotope in soils to identify the source of lead contamination near Adelaide, South Australia. Environ. Sci. Technol. 15(6):691-696.

Gutiérrez, M., K. Mickus, and L.M. Camacho. 2016. Abandoned Pb–Zn mining wastes and their mobility as a proxy to toxicity: A review. Sci. Total Environ. 565:392-400.

Gwiazda, R.H., and D.R. Smith. 2000. Lead isotopes as a supplementary tool in the routine evaluation of household lead hazards. Environ. Health Perspect. 108(11):1091-1097.

Hallett, R.L., and C.H. Rose. 1935. Lead pigments. Pp. 81-87 in Symposium on Paint and Paint Materials. STP38728S. West Conshohocken, PA: ASTM International [online]. Available: https://doi.org/10.1520/STP38728S [accessed May 22, 2017].

Jacks, G., M. Byström, and L. Johansson. 2001. Lead emissions from lost fishing sinkers. Boreal Environ. Res. 6:231-236.

Johnson, C.E., A.H. Johnson, T.G. Huntington, and T.G. Siccama. 1991. Whole-tree clear-cutting effects on soil horizons and organic-matter pools. Soil Sci. Soc. Am. J. 55:497-502.

Johnson, C.E., T.G. Siccama, C.T. Driscoll, G.E. Likens, and R.E. Moeller. 1995. Changes in forest lead cycling in response to decreasing atmospheric inputs. Ecol. Appl. 5(3):813-822.

Johnson, C.E., R.J. Petras, R.H. April, and T.G. Siccama. 2004. Post-glacial lead dynamics in a forest soil. Water Air Soil Pollut. Focus 4(2):579-590.

Kaiser, C.J., W.C. Kelly, R.J. Wagner, and W.C. Shanks. 1987. Geologic and geochemical controls of mineralization in the Southeast Missouri barite district. Econ. Geol. 82(3):719-734.

Kendall, R.J., T.E. Lacker. Jr., C. Bunck, B. Daniel, C. Driver, C.E. Grue, F. Leighton, W. Stansley, P.G. Watanabe, and M. Whitworth. 1996. An ecological risk assessment of lead shot exposure in non-waterfowl avian species: Upland game birds and raptors. Environ. Toxicol Chem. 15(1):4-20.

Leach, D.L., D.F. Sangster, K.D. Kelley, R.R. Large, G. Garven, C.R. Allen, J. Gutzmer, and S. Walters. 2005. Sediment-hosted lead-zinc deposits: A global perspective. Pp. 561-607 in Economic Geology: One Hundredth Anniversary Volume 1905–2005, J.W. Hedenquist, J.F.H. Thompson, R.J. Goldfarb, and J.P. Richards, eds. Littleton, CO: Society of Economic Geologists.

Leach, D.L., R.D. Taylor, D.L. Fey, S.F. Diehl, and R.W. Saltus. 2010. A Deposit Model for Mississippi Valley-Type Lead-Zinc Ores, Chapter A of Mineral Deposit Models for Resource Assessment. U.S. Geological Survey Scientific Investigations Report 2010–5070–A [online]. Available: https://pubs.usgs.gov/sir/2010/5070/a/ [accessed September 18, 2017].

Long, K.R., J.H. DeYoung, and S.D. Ludington. 1998. Database of Significant Deposits of Gold, Silver, Copper, Lead, and Zinc in the United States. U.S. Geological Survey Open-File Report 98-206. Reston, VA: U.S. Geological Survey.

Manton, W.I. 1985. Total contribution of airborne lead to blood lead. Br. J. Ind. Med. 42(3):168-172.

McHenry, R.E. 2006. Chat Dumps of the Missouri Lead Belt, St. Francois County, With an Illustrated History of the Lead Companies that Built Them. Park Hills, MO: Missouri Department of Natural Resources, Missouri Mines State Historic Site.

McLimans, R.K., H.L. Barnes, and H. Ohmoto. 1980. Sphalerite stratigraphy of the upper Mississippi Valley zinc-lead district, southwest Wisconsin. Econ. Geol. 75(3):351-361.

Mielke, H.W. 1994. Lead in New Orleans soils: New images of an urban environment. Environ. Geochem. Health 16(3):123-128.

Mielke, H.W. 1999. Lead in the inner cities: Policies to reduce children's exposure to lead may be overlooking a major source of lead in the environment. Am. Sci. 87(1):62-73.

Mielke, H.W., and P.L. Reagan. 1998. Soil is an important pathway of human lead exposure. Environ. Health Perspect. 106(Suppl. l):217-229.

MO DNR (Missouri Department of Natural Resources). 2017. Missouri Lead Mining History by County. Missouri Department of Natural Resources [online]. Available: https://dnr.mo.gov/env/hwp/sfund/lead-mo-history-more.htm [accessed February 15, 2017].

Mugel, D.N. 2017. Geology and Mining History of the Southeast Missouri Barite District and the Valles Mines, Washington, Jefferson, and St. Francois Counties, Missouri. U.S. Geological Survey Scientific Investigations Report 2016-5173 [online]. Available: https://pubs.usgs.gov/sir/2016/5173/sir20165173.pdf [accessed June 20, 2017].

Murakami, M., F. Nakajima, H. Furumai, B. Tomiyasu, and M. Owari. 2007. Identification of particles containing chromium and lead in road dust and soakaway sediment by electron probe microanalyser. Chemosphere 67(10):2000-2010.

Papp, J.F. 2013. Recycling metals. Pp. 61.1–61.5 in Minerals Yearbook 2013, Vol. 1. Metals and Minerals [online]. Available: https://minerals.usgs.gov/minerals/pubs/commodity/recycle/myb1-2013-recyc.pdf [accessed June 20, 2017].

Peryea, F.J. 1998. Historical use of lead arsenate insecticides, resulting soil contamination and implications for soil remediation. Scientific Registration No. 274 in Proceedings of the 16th World Congress of Soil Science,

August 20-26, 1998, Montpellier, France [online]. Available: http://soils.tfrec.wsu.edu/leadhistory.htm [accessed May 22, 2017].

Plumlee, G.S. 1999. The environmental geology of mineral deposits. Pp. 71-116 in The Environmental Geochemistry of Mineral Deposits, Part A. Processes, Techniques, and Health Issues. Review of Economic Geology Vol. 6A. Littleton, CO: Society of Economic Geologists.

Ramos, F.C., and P.E. Rosenberg. 2012. Age and origin of quartz-carbonate veins associated with the Coeur d'Alene mining district, Idaho and western Montana: Insights from isotopes and rare-earth elements. Econ. Geol. 107(6):1321-1339.

Rao, V., L. Tooly, and J. Drukenbrod. 2013. 2008 National Emissions Inventory: Review, Analysis, Highlights. EPA-454/R-13-005. U.S. Environmental Protection Agency, Office of Air Quality Planning and Standards, Research Triangle Park, NC [online]. Available: https://www.epa.gov/sites/production/files/2015-07/documents/2008report.pdf [accessed June 20, 2017].

Ridley, J. 2013. Ore Deposit Geology. New York: Cambridge University Press.

Rose, A.W., H.E. Hawkes, and J.S. Webb. 1979. Geochemistry in Mineral Exploration. London: Academic Press. 657 pp.

Rudnick, R.L., and S. Gao. 2014. Composition of the continental crust. Pp. 1-51 in Treatise on Geochemistry, Vol. 4. The Crust, 2nd Ed., H. Holland, and K. Turekian, eds. Amsterdam: Elsevier.

Sahl, K., B.R. Doe, and K.H. Wedepohl. 1978. Lead. Pp. 80A-1 to 82-O-1 in Handbook of Geochemistry Vol. 2, K.H. Wedepohl, ed. New York: Springer.

Sangster, D.F., P.M. Outridge, and W.J. Davis. 2000. Stable lead isotope characteristics of lead ore deposits of environmental significance. Environ. Rev. 8(2):115-147.

Scheuhammer, A.M., S.L. Money, D.A. Kirk, and G. Donaldson. 2003. Lead Fishing Sinkers and Jigs in Canada: Review of Their Use Patterns and Toxic Impacts on Wildlife. Canadian Wildlife Service Occasional Paper 108. Ottawa, Canada: Environment Canada. 48pp.

Schooley, T., M.J. Weaver, D. Mullins, and M. Eick. 2008. The history of lead arsenate use in apple production: Comparison of its impact in Virginia with other states. J. Pest. Saf. Educ. 10:22-53.

Smith, G.R. 2004. Lead recycling in the United States in 1998. Chapter F in Flow Studies for Recycling Metal Commodities in the United States, S.F. Sibley, ed. U.S. Geological Survey Circular 1196. Reston, VA: U.S. Geological Survey [online]. Available: https://pubs.usgs.gov/circ/2004/1196am/c1196a-m_v2.pdf [accessed June 20, 2017].

Smith, K.S. 2007. Strategies to predict metal mobility in surficial mining environments. Pp. 25-45 in Understanding and Responding to Hazardous Substances at Mine Sites in the Western United States, J.V. DeGraff, ed. Reviews in Engineering Geology Vol. 17. Boulder, CO: Geological Society of America [online]. Available: https://minerals.usgs.gov/east/mea/Smith2007_508.pdf [accessed June 20, 2017].

Smith, K.S, and H.L.O. Huyck. 1999. An overview of the abundance, relative mobility, bioavailability, and human toxicity of metals. Pp. 29-70 in The Environmental Geochemistry of Mineral Deposits, Part A: Process, Techniques and Health Issue. Littleton, CO: Society of Economic Geologists.

Smith, W.H. 1976. Lead contamination of the roadside ecosystem. J. Air Pollut. Control Assoc. 26(8):753-766.

Spinosa, E.D., D.T. Hooie, and R.B. Bennett. 1979. Summary Report on Emissions from the Glass Manufacturing Industry. EPA/600/2-79/101. U.S. Environmental Protection Agency, Industrial Environmental Research Laboratory, Cincinnati, OH.

USGS (U.S. Geological Survey). 2015. Mineral Resources Data System (MRDS) [online]. Available: https://mrdata.usgs.gov/mrds/ [accessed February 2, 2017].

Vignes, A. 2013. Extractive Metallurgy, 3. Processing Operations and Routes. London: iSTE.

Wharton, H.M. 1972. Barite Ore Potential of Four Tailings Ponds in the Washington County Barite District, Missouri. Rolla, MO: Geological Survey and Water Resources.

Wilburn, D.R. 2014. Comparison of the U.S. Lead Recycling Industry in 1998 and 2011. U.S. Geological Survey Scientific Investigations Report 2014-5086. Reston, VA: U.S. Geological Survey [online]. Available: https://pubs.usgs.gov/sir/2014/5086/pdf/sir2014-5086.pdf [accessed June 20, 2017].

Winslow, A. 1894. Lead and Zinc Deposits. Jefferson City, MO: Missouri Bureau of Geology and Mines.

Zientek, M.L., and G.J. Orris. 2005. Geology and Nonfuel Mineral Deposits of the United States. U.S. Geological Survey: Open-File Report 2005-1294-A. Reston, VA: U.S. Geological Survey [online]. Available: https://pubs.usgs.gov/of/2005/1294/a/of2005-1294a.pdf [accessed September 17, 2017].

3

Environmental Dispersal of Lead

Chapter 2 described many possible sources of lead that could cause environmental contamination. To assess whether a lead source might be contributing to observed contamination, one must first consider possible dispersal pathways. As shown in Figure 2-1, there are natural and intentional transport mechanisms; in this chapter, the committee discusses those mechanisms and how they might promote the dispersal of lead into the environment. Aqueous processes and transport are described first, then air transport, and finally intentional transport.

AQUEOUS PROCESSES AND TRANSPORT

A primary natural transport mechanism is water. In this section, the committee discusses lead solubility and species in water, lead partitioning between water and solid phases, and sediment-associated dispersal in rivers.

Lead Solubility and Species in Water

Dissolved lead concentrations can vary by orders of magnitude but are generally lower than those of other dissolved species. In uncontaminated fresh and saline waters, concentrations typically range from around 0.2 to 200 ppt (about 10^{-12}–10^{-9} mol/L). In contaminated waters, lead concentrations typically rise into the range of a few hundred parts per billion (about 10^{-6} mol/L). For example, a river contaminated by acid mine drainage and dissolved metals had a maximum lead concentration of 698 ppb (3.4×10^{-6} mol/L) and a mean concentration of 121 ppb measured over 2 years (Nieto et al. 2007); in contrast, mine runoff and discharge waters at near-neutral pH in a lead mining area had lead concentrations that averaged 2.6–14 ppb over 1.5 years (Schaider et al. 2014). For comparison, in a 7-month period in 1966-1967, the mean total lead concentration in rainfall in the United States was 34 ppb, and lead concentrations were higher near cities because of emissions from industrial activities and gasoline (Lovering 1976). As a result of lead accumulation in soils related to lead-containing precipitation before 1976, lead concentrations as high as 330 ppb were measured in runoff water from soils in some parts of New England (Hem 1976). In the United States, emissions of lead to the atmosphere and later deposition on land decreased substantially after the early 1980s, when the addition of lead to gasoline was phased out (Callender 2005). In highly contaminated surface waters and groundwater, lead concentrations can exceed 1,000 ppb (or 1 ppm), but this is generally rare, and most waters with high dissolved lead have concentrations well below parts per million. Away from contamination sources, lead is typically reduced to low concentrations by mixing and dilution with uncontaminated waters and by adsorbing to particles in the water column as described in the next section.

In most surface environmental settings, the lead in water is predominantly divalent lead, Pb^{2+}. In the absence of dissolved organic matter, Pb^{2+} has a strong tendency to form hydroxo species, such as $PbOH^+$, in solution with increasing pH (Figure 3-1A). In common surface waters and groundwater, the dissolved carbonate is typically about 10^{-5}–10^{-3} mol/L total carbonate, and lead-carbonate complexes will dominate over lead-hydroxo species over the typical pH range of natural waters, about 6.5–9.5 (Figure 3-1B). In contaminated waters that contain high dissolved sulfate, such as those associated with acid mine drainage, free Pb^{2+} is the dominant species below a pH of about 7, and the aqueous complex with sulfate, $PbSO_4$(aq), is a minor species (Powell et al. 2009). In seawater and high dissolved-chloride systems, lead-chloride complexes—$PbCl^+$, $PbCl_3^-$, and $PbCl_2$(aq)—dominate over free Pb^{2+} below a pH of about 7.5. At alkaline pH and high chloride, dissolved lead is present mostly as aqueous lead-carbonate complexes, $PbCO_3$(aq) and $Pb(CO_3)_2^{2-}$, rather than free Pb^{2+} or mixed carbonate-chloride and carbonate-hydroxo complexes (Powell et al. 2009).

Dissolved organic matter (DOM) forms strong complexes with Pb^{2+} that can dominate speciation in water, depending on the DOM concentration (Town and Fillela 2000a,b; Rozan et al. 2003). Because natural DOM is a heterogeneous and complex mixture that contains different organic functional groups, many binding-site concentrations and stability constants for lead complexation have been reported (Town and Fillela 2000b). Researchers have attempted to use simple organic acids and ligands as proxies for DOM to describe lead–DOM interactions, but those attempts have had mixed success (Rozan et al. 2003; Vega and Weng 2013; Stockdale et al. 2015).

As discussed in the next section, lead can also be transported in surface waters or groundwater in association with colloidal or suspended particles. After a water sample is collected, it might or might not be filtered before the lead concentration is analyzed. If filtered, the pore size of the filter used can influence the measured lead concentration by removing fine particles that contain lead. A standard filter pore size is 0.45 µm (EPA 1996). The lead concentration measured in a sample after passing through a 0.45-µm filter is typically referred to as the dissolved concentration, whereas *total lead* generally refers to lead measured in unfiltered water samples after acid digestion. The two measurements differ if filtration removes particles that contain lead larger than the filter pore diameter. For example, in one study of more than 100 groundwater samples from private domestic water wells in Pennsylvania, total (unfiltered) lead concentrations varied from 0 to 100 ppb. When dissolved (filtered) lead concentrations were compared with total lead in the water samples, the authors determined that dissolved lead concentrations were lower by 6–18 ppb (Swistock et al. 1993).

Lead Partitioning

Given the low concentrations of dissolved lead found in most natural waters, surface sorption (adsorption or ion exchange) of Pb^{2+} onto mineral surfaces or organic matter will likely be the most common mechanism of lead removal from solution to solid phases in surface environments rather than precipitation of new lead species. Precipitation of Pb^{2+} salts, however, occurs in the surface weathering of lead minerals, in evaporative scenarios, and from high-lead surface waters or groundwater near, for example, mining sites or industrial lead sources (see below). Sorption of lead from solution onto mineral surfaces is controlled by pH, solution concentration, and type of mineral surface (Stumm 1992). For example, dissolved lead has a stronger binding affinity to iron hydroxide mineral surfaces than to aluminum hydroxide surfaces, and this results in greater lead adsorption to surfaces of iron hydroxide than to aluminum hydroxide at lower pHs if

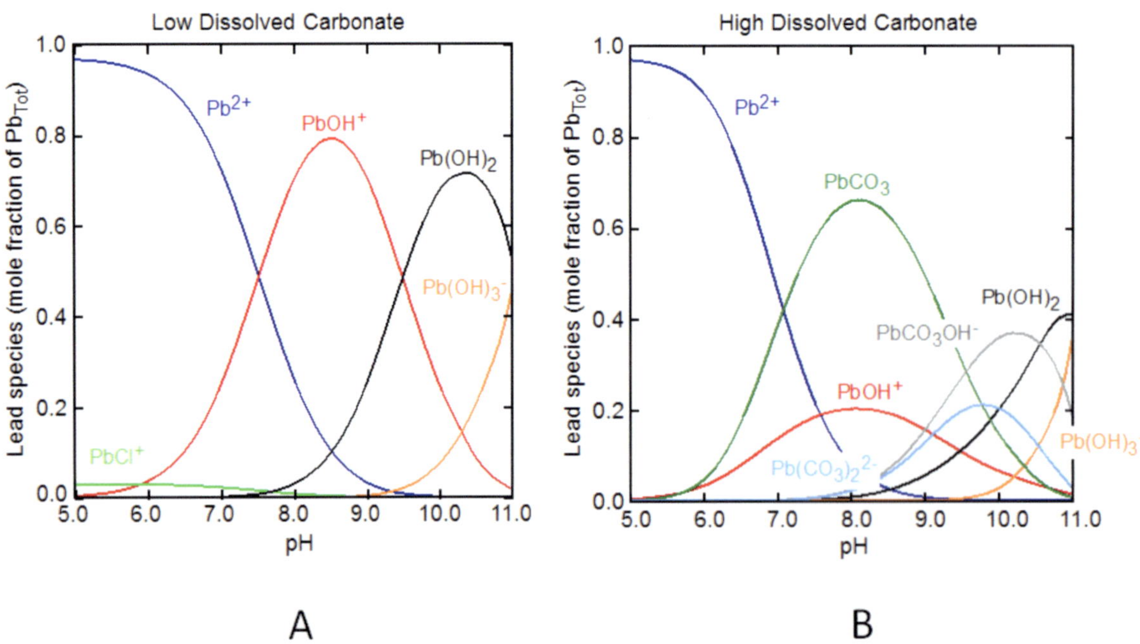

FIGURE 3-1 Dominant lead species in water as a function of pH at (A) low and (B) high dissolved-total-carbonate concentrations. *Low* dissolved carbonate refers to a solution in equilibrium with a partial pressure of CO_2 (g) (P_{CO2}) of about 10^{-5} atm, and *high* dissolved carbonate refers to a solution in equilibrium with a P_{CO2} of about $10^{-3.5}$ atm (about atmospheric CO_2). Calculated with PhreeqC by using equilibrium constants from Powell et al. (2009). Lead species that constitute less than about 1% of the total lead concentration of 2×10^{-6} mol/L are not shown. Source: Adapted from Powell 2009. Reprinted with permission; copyright 1999, *Pure and Applied Chemistry*.

the solution concentration and mineral surface area are the same (Figure 3-2) (Serrano et al. 2009). The high affinity of lead for iron and manganese oxides also allows it to outcompete other ions for sorption. Clay minerals can bind Pb^{2+}, particularly at acidic pH, but the bound lead can be displaced by cations, such as Ca^{2+}, when they are present at concentrations similar to that of the Pb^{2+}. Pb^{2+} sorbs to calcite, $CaCO_3$, and other carbonate minerals at alkaline pH (Rouff et al. 2005).

Sorption of Pb^{2+} on mineral surfaces is relatively fast, typically minutes to hours. Depending on the mineral type and aging time, desorption of lead can be mostly reversible (that is, lead is completely removed from the surface and goes back into solution) or only partly reversible (that is, some lead remains bound to the surface) if pH or solution conditions change. Longer reaction or aging time can lead to incorporation of Pb^{2+} into a mineral structure or irreversible binding. For example, a minor amount of Pb^{2+} might be substituted for the divalent cation in carbonate mineral phases, or it might be incorporated into the structure of hydrated iron oxide phases (ferrihydrite or hydrous ferric oxide); this requires dissolution of the solid phase to release lead into solution. The observed slow release of lead from soils in leaching experiments has been attributed to the strong binding of Pb^{2+} to oxoanions in the interlayer of clay minerals and the resulting low rates of desorption from those sites (Serrano et al. 2013).

In soil and sediment weathering, lead-ore minerals, such as galena (PbS), might oxidize, dissolve, or change to form new minerals that might (or might not) be lead-rich solids, such as mixed-element salts or mineral solid solutions. The amount of weathering, rainfall, and leaching in a given environment determines whether lead minerals dissolve completely, change to other solids, or adsorb to other mineral surfaces if dissolved (Hayes et al. 2009). For example, in environments with high rainfall, such ions as lead tend to leach from soils to water, where they are transported as dissolved species, or to adsorb to particles or organic matter in water and migrate as particle-associated species (Schaider et al. 2014; Gutierrez et al. 2016). Organic matter in soils, which is typically more abundant in temperate and tropical environments, tends to retain lead by adsorption and hinder its release to water (Marzouk et al. 2013). In contrast, in arid and semiarid regions where soils undergo intermittent wetting and drying, weathering might result in the formation of carbonate, clay, hydrous oxides, sulfate, or mixed sulfate–hydroxide phases that incorporate lead into their structure or adsorb lead on their surfaces (Dutrizac and Jambor 2000; Jambor et al. 2000).

The pH of water—either surface water or water in the pores of soils and sediments—exerts a strong influence on the dissolution of lead solids derived from original sources (primary phases), the formation of new lead-

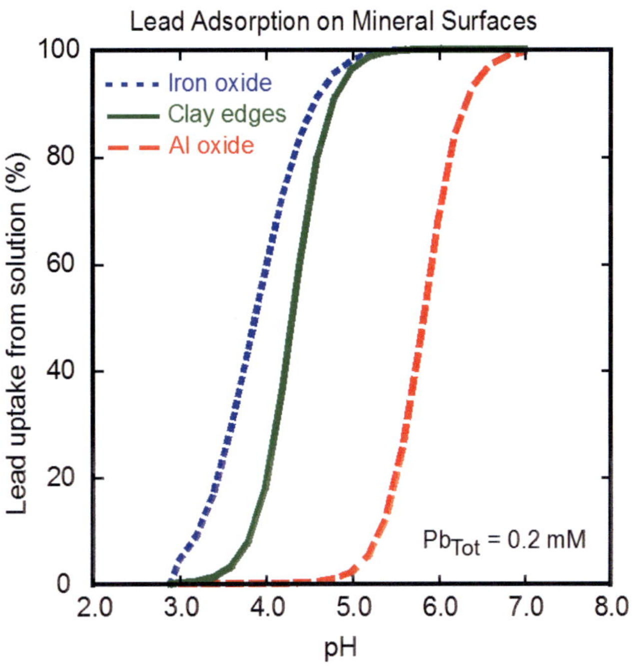

FIGURE 3-2 General examples of lead adsorption (as percent uptake from solution) on mineral surfaces as a function of pH for three soil-mineral types: iron oxide or hydroxide, edge sites of clay minerals, and aluminum oxide or hydroxide. The pH range over which lead sorbs in real systems will vary with solution and mineral-surface-site concentrations. Calculated with PHREEQC by using sorption parameters from Serrano et al. (2009). Source: Data adapted from Serrano et al 2009. Reprinted with permission; copyright 2009, *Geochimica et Cosmochimica Acta*.

containing solids by precipitation (secondary phases), and lead adsorption to the surfaces of other soil minerals. The presence of carbonate minerals, such as calcite, in soils and sediments buffers pH near neutral or above (about 7.2–8.2). Given those conditions, lead is often retained by the precipitation of new carbonate minerals or by adsorption. For example, cerussite, $PbCO_3$, was noted as a major secondary weathering phase in sediments that were affected by a historic mining district of Wanlockhead, Scotland (Hillier 2001). Studies of the weathering of lead munitions illustrated that cerussite and hydrocerussite, $Pb_3(CO_3)_2(OH)_2$, were primary products that likely control dissolved lead concentrations in soil-pore water (Jørgensen and Willems 1987; Lin et al. 1995; Cao et al. 2003a; Hardison et al. 2004; Rooney et al. 2007). Similarly, formation of cerussite and mixed lead hydroxide–carbonate phases—including hydrocerussite and plumbonacrite, $Pb_5(CO_3)_3O(OH)_2$—have been reported as products of corrosion of lead pipes (Xie and Giammar 2011; Masters et al. 2016) or products of direct weathering of galena at alkaline pH (Hudson-Edwards et al. 1996; Root et al. 2015). Soils or sediments that have high phosphate concentrations, such as agricultural soils that receive high concentrations of fertilizer, might contain pyromorphite, $Pb_5(PO_4)_3Cl$ (Cao et al. 2003b). As a result of weathering or alteration, secondary lead phases form as crusts or scales on surfaces of galena, lead shot, or lead pipe where lead concentrations are relatively high (Cao et al. 2003a; Pieper et al. 2017). However, at lower concentrations as lead is dispersed, adsorption on clay minerals or amorphous iron oxides or aluminum oxides in soils and sediments is often the primary mechanism for retaining lead and limiting its mobility in water (Hillier 2001; Yang et al. 2006; Rooney et al. 2007).

In areas that have lead-bearing ore minerals and wastes from deposits that contain abundant iron sulfide minerals that produce sulfuric acid during weathering (see Chapter 2), extensive weathering and leaching result in progressively more acidic waters as carbonate is removed and buffer capacity is lost (Hayes et al. 2009; Nordstrom et al. 2015). In acidic conditions in soils and water, the concentrations of iron, aluminum, and sulfate are important controls on lead dispersal or retention in surface weathering or evaporative scenarios. Assemblages of anglesite, $PbSO_4$, and plumbojarosite, $PbFe_6(SO_4)_4(OH)_{12}$, with a wide variety of mixed-metal salts that contain iron, aluminum, and sulfate are common products of lead-rich weathered ore deposits and mine wastes (Dutrizac and Jambor 2000; Jambor et al. 2000). Soils and evaporated sediments that have high dissolved lead and sulfate concentrations, such as those associated with tailings impoundments or mine drainage, might contain anglesite; possibly linarite, $PbCu(SO_4)(OH)_2(s)$, if copper concentrations are high; or lead substituted in many other hydrated-sulfate salts (Jørgensen and Willems 1987; Bigham and Nordstrom 2000; Jambor et al. 2000). The sulfate–hydroxide phases often precipitate and dissolve quickly as conditions change, but reaction rates in different environments will depend on particle size, mineral crystallinity, pH, and solution composition.

Sediment-Associated Lead Dispersal in Rivers

As noted above, lead has a high propensity to be sorbed (attached) to particulate matter under the geochemical conditions observed in most rivers. The concentration of lead associated with particulate matter also tends to be orders of magnitude higher than the concentrations of dissolved lead in surface water or groundwater. Given the concentrations of lead associated with particles and the ability of rivers to move large quantities of sediment, rivers are highly efficient in transporting lead considerable distances downstream from a natural or anthropogenic source. The movement of sediment-associated lead from historical mining and milling operations can be particularly pronounced because waste materials were often stored within impounded river valleys and allowed to enter a river channel unabated.

The downstream transport of sediment-associated lead is a function of the size of the sediment that is being moved and the energy provided by the flow to perform mechanical work. In general, the higher the flow (discharge), the more sediment and sediment-associated lead can be transported through a given stream segment, although the relationship between flow and sediment transport is complicated. Transport of suspended sediment depends on factors—such as the existing soil moisture conditions; rainfall volume, duration, and intensity; and vegetation cover—that influence the movement of sediment from upland areas to the river. As a result, concentrations of suspended sediment and sediment-associated lead vary widely between floods and nonlinearly during a single event, and quantification of suspended sediment and particulate lead transported through a stream segment requires the monitoring of numerous floods over multiple years and an understanding of how concentrations of particulate lead change during a runoff event.[1]

The downstream dispersal of sediment-associated lead (and other trace metals) from a point source generally creates a distinctive systematic downstream decrease in lead concentrations in the sediments in the channel and its associated valley. Figure 3-3 provides an example of the spatial trends commonly encountered for metals downstream of point sources of contamination along the Rio Chilco–Rio Tupiza drainage system of southern Bolivia. Lead, zinc, and antimony concentrations are relatively low at site RC-3 and are comparable with local background values. Concentrations of the metals or metalloids abrupt-

[1]See Richards (2001) for a summary of methods to estimate pollutant loads.

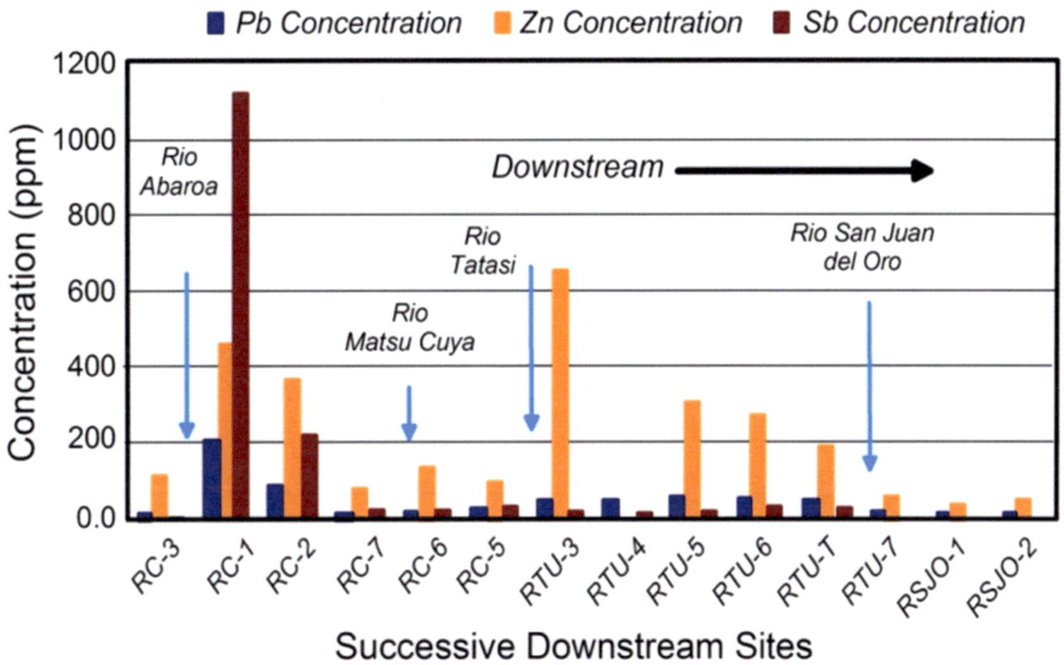

FIGURE 3-3 Lead, zinc, and antimony concentrations measured in channel-bed sediments of the Chilco–Tupiza drainage system of southern Bolivia. Antimony mines are in the Rio Abaróa and Rio Matsu Cuya tributary basins. Polymetallic tin deposits that contain little antimony are in the Tatasi River basin. Point sources of contamination are indicated by abrupt increases in element concentration followed by downstream declines. Source: Villarroel et al. (2006). Reprinted with permission; copyright 2006, *Environmental Geology*.

ly increase immediately downstream of the Rio Abaróa, a tributary contaminated by antimony mining and a tailings impoundment failure (Villarroel et al. 2006). The abrupt increase is presumed to result from the influx of contaminated sediment from the mine site. Concentrations of all three elements then decline systematically downstream. However, zinc concentrations abruptly increase again on reaching the confluence of the Rio Tatasi and to a lesser degree the Rio Matsu Cuya. Both drainages contain large-scale mining operations where tailings have entered the channel. Thus, the increased concentrations reflect the additional input of zinc-contaminated tailings into the river.

Longitudinal decreases in lead concentrations, such as those documented downstream of the Rio Abaróa, result from the interaction of multiple processes (Figure 3-4) (Lewin and Macklin 1987; Macklin 1996; Miller and Orbock Miller 2007). Those processes include (1) dilution associated with the dispersal of sediment-associated lead from a relatively small source area over a much longer segment of the river, (2) dilution caused by the mixing of contaminated particles with clean sediment from tributaries and eroded bank deposits, (3) the deposition and storage of lead in the channel bed and other alluvial deposits (a process that removes sediment-associated lead from further transport), and (4) the hydraulic sorting of contaminated particles as a result of the selective erosion, transport, and deposition of larger or denser grains, such as galena. In addition, geographic patterns in lead concentrations can be influenced by the geochemical processes described earlier. Although the downstream dispersal of sediment-associated lead can be systematic, a number of processes complicate geographic patterns in dispersal. The most important are described below.

Lead Partitioning between Particles That Have Different Physical Properties

In most rivers, lead is largely associated with chemically reactive, fine-grained sediments (Horowitz 1991). During their downstream transport, the sediments might be hydraulically separated by particle size and density in localized areas of the riverbed and its floodplain; each area has a distinctive range of particle sizes. Such areas in the river include pools, riffles, point bars, and glides (Figure 3-4a). Within the floodplain or bank, the particles form packages of sediment referred to as vertical accretion deposits, lateral accretion deposits, crevasse-splay deposits, and oxbow lake deposits, and others (Figure

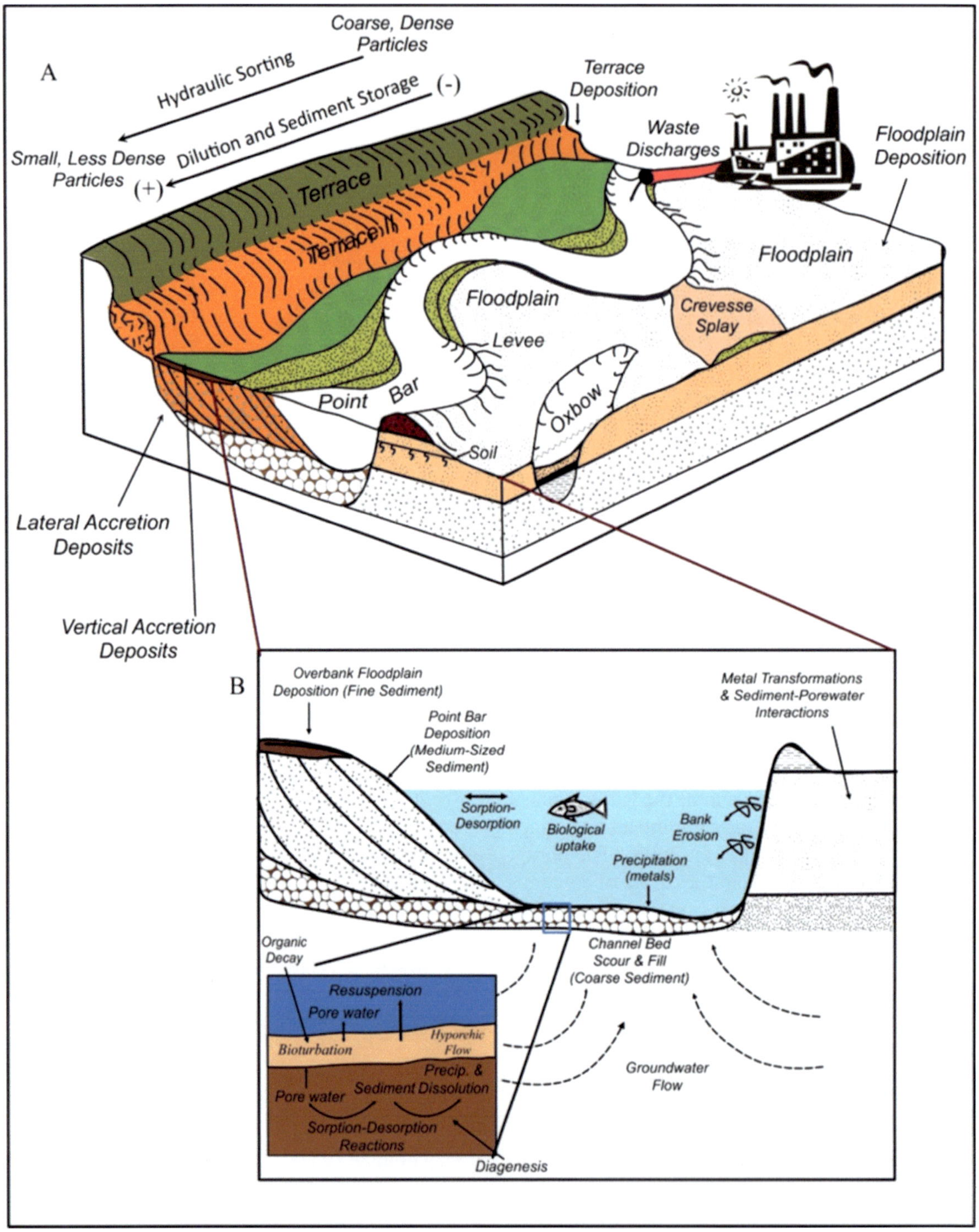

FIGURE 3-4 Schematic diagrams that illustrate potential sites of sediment-associated lead deposition in a river and the primary factors that influence lead concentrations in channel-bed sediment. (A) Processes operating along a stream segment include hydraulic sorting, dilution, and sediment storage; partition of sediment by size and density into various morphologic units is also shown. (B) Geochemical and biologic processes—including oxidation–reduction, precipitation, dissolution, and cation exchange processes—might also influence geographic patterns of contaminant concentration. Source: Adapted from Miller and Orbock Miller (2007). Reprinted with permission; copyright 2007, Springer.

3-4a). Such particle partitioning results in the concentration of lead in specific areas of the drainage network that contain fine-grained, lead-enriched particles (Macklin et al. 1999; Macklin et al. 2006; Miller and Orbock Miller 2007).

Along rivers that are affected by the influx of mining and milling debris, lead is associated not only with fine-grained sediment but with coarse-grained sediment (Moore et al. 1989; Miller et al. 1998). The exact distribution depends largely on the milling process used and on how metals were incorporated into the tailings (waste) materials. For example, Pavlowsky et al. (2010) found that high lead concentrations in tailings from the Leadwood and National Piles in Southeast Missouri occurred in the sediment fraction measuring less than 63 µm and in the fraction measuring 4–8 mm. Lead concentrations associated with particles of an intermediate size were relatively low because the milling process was more efficient in recovering lead (and zinc) from these particle sizes. Thus, the geographic distribution of sediment-associated lead is likely to reflect the spatial distribution of particle sizes in the river valley.

In addition to particle size, grain density might affect partitioning of sediment, particularly in mineralized areas. For example, some minerals, such as gold and galena, have a relatively high density. Those dense mineral grains are often concentrated in localized zones in which the denser but smaller metallic grains coexist with larger but less dense minerals, such as quartz, feldspar, and carbonate minerals. The zones of concentration, which are called placer deposits, are created by the repeated processes of erosion and redeposition of particles of a given size range (Slingerland and Smith 1986). Placers tend to form upstream of natural protrusions that are caused by resistant strata where bedrock is present and upstream of manmade structures, such as road crossings and small dams. Placers might also form along rivers where the channel perimeter is composed of sediments. Those types of placers are formed in areas where the repeated erosion and redeposition of sediment concentrates the denser particles, such as at the base of point bars, the head of midchannel bars or islands, and the confluence of tributaries (Guilbert and Park 1986).

Not all differences in the concentration of sediment-associated metals observed between the depositional units of a river can be attributed to particle size and density (Graf et al. 1991; Hudson-Edwards et al. 2001). Areas in the channel that are inundated or flooded more often commonly exhibit higher sediment-associated metal concentrations (Graf et al. 1991). The influence of inundation frequency on metal concentrations is related to two primary factors. First, the more an area of the channel is flooded, the greater the opportunity for sediment-associated lead to be incorporated into the area's channel sediments. Second, the frequency of inundation depends on flood size: topographically higher and less frequently inundated areas are flooded during larger runoff events. In many contaminated rivers, sediment-associated metal concentrations decrease with increasing flow because there is an increased influx of clean sediments to the channel from upland areas of the basin. The clean sediments decrease the concentrations of the metals in the sediments that are incorporated into the less frequently inundated channel areas (Marron 1987).

Localized Downstream Variations in Sediment-Associated Lead Transport and Deposition

Many river systems exhibit abrupt changes in their capacity to transport sediment along the drainage network (Macklin and Lewin 1989; Miller and Villarroel 2011; Miller et al. 2012). For example, segments of the stream characterized by higher gradients and narrower than average valleys tend to transport sediment and sediment-associated lead with little deposition. In essence, flood waters are confined by the narrowness of the valley to a smaller area, which when combined with higher gradients creates more energetic and erosive flows. In contrast, sediment transport is reduced in sedimentation zones that are characterized by wide valleys and shallowness (Mackin and Lewin 1989; Mackin and Dowsett 1989; Mackin 1996; Miller et al. 2009). As a result, systematic downstream trends in the concentration and storage (deposition) of sediment-associated lead might be locally disrupted by changes in river and valley gradient and width. The influence of river or valley gradient and width on geographic patterns of sediment-associated metal concentrations is most pronounced along rivers in mountainous terrain but can also be important along lowland rivers.

Influence of River Transformations on Sediment-Associated Lead Dispersal

Lewin and Macklin (1987) argued that the dispersal of sediment-associated metals differs between rivers in a stable, equilibrium state and those in an unstable, disequilibrium state. The latter are characterized by continuing changes or transformations in channel width, depth, slope, and sinuosity caused by localized erosion or deposition. Transformations in river structure are important because they influence the rate at which sediment-associated metals are transported to downstream areas and are important determinants of the sites along the river where sediment-associated metals are deposited and stored. A relevant example is the influx of large quantities of sediment and tailings into a river from historical mining or milling operations. The increased sediment load often overwhelmed the transport capacity of the river near the site of sediment input and caused deposition on the riverbed and on the adjacent floodplain. Deposition was often substantial and in some cases measured several feet to tens of feet in thickness (Gilbert 1917; James 1989; Knighton 1991). Once the influx of sediment from the mines or mills decreased, the river was able to transport the previously deposited sediment downstream, but some remained in the stream segment and was exposed within the riverbanks, where it could be eroded and re-enter the river. The sediment that was transported downstream often overwhelmed that portion of the river and caused deposition there until sediment loads declined, at which point the riverbed was eroded and the cycle repeated itself further downstream. When viewed in its entirety, the cycle of deposition and erosion formed a kind of sediment wave that migrated downstream from the site of sediment input (Gilbert 1917; Knighton 1991; Nicholas et al. 1995). The downstream passage of the wave typically took decades to complete and strongly influenced the rate at which sediment-associated metals were transported along the river.

The influx of sediment from historical mining and milling operations is not the only type of disturbance that can cause such transformations in river structure and sediment transport rates. In fact, such changes in river function are associated with a wide array of natural disturbances (such as landslides and wildfires) and anthropogenic activities (such as deforestation and land-use change). One of the most widespread and extensively documented disturbances in the United States that led to such transformations in river form was land-use change initiated by European settlement during the late 19th and early 20th centuries (Trimble 1974; Knox 1989; Orbock Miller et al. 1993; Price and Leigh 2006; Leigh 2010). Specifics of the documented changes in the river basins vary between sites; however, they were generally characterized by (1) large-scale devegetation of hillslope and valley bottoms that resulted in an increase in runoff to the drainage network, (2) gullying and accelerated erosion of upland soils that led to an increase in sediment loads into the channel, (3) deposition of sediment on floodplains or riverbeds that produced what are commonly called legacy sediments, and (4) a period of reforestation and improved watershed management practices that resulted in reduced runoff, sediment yield, and, in many cases, channel erosion (Trimble 1974; Knox 1987).

During the above transformations in rivers and river valleys, lead associated with natural soil particles might be incorporated into the sediments that are deposited along the river, particularly the floodplain. After erosion of the riverbed, those sediments (including any lead that they contain) will be exposed in the riverbanks (see Figure 3-5), where they might continue to be eroded and reintroduced to the river (James 1989; Miller et al. 1998; Bain et al. 2012; Niemitz et al. 2013). Such processes might be particularly important along rivers that drain mineralized rocks, such as in Southeast Missouri.

FIGURE 3-5 Mine tailings incorporated into floodplain deposits near Leadville, Colorado. Erosion reintroduces sediment-associated lead into the drainage network.

The erosion and lowering of the river bed also forms an alluvial terrace. Terraces are old floodplains that because of erosion and lowering of the river bed now exist topographically above and adjacent to the river and its floodplain (Figure 3-4). Terraces often contain sediment-associated lead that was deposited by the river but is now spatially removed from the modern channel. Geographic patterns in sediment-associated lead concentrations in terraces are influenced by the timing of lead inputs into the river, terrace height above the modern channel, and the frequency with which the terraces are flooded. During the period of lead influx, lower elevation terraces are likely to have higher concentrations and larger quantities because they are inundated more frequently. After the influx of lead, contaminated sediments on the lower terraces are likely to be buried more rapidly by sediments that have lower lead concentrations. Thus, lead-contaminated soils can be at or closer to the surface on higher terraces than on lower terraces or the floodplain (Brewer and Taylor 1997).

AIR TRANSPORT

Airborne transport has the greatest potential to disperse lead emissions because the air is not confined by geographic barriers. Furthermore, rates of transport of contaminants are often higher for the air pathway than for groundwater and surface-water pathways (Prabhakar et al. 2014). The potential for lead transport via the air pathway can be illustrated with data from Southeast Missouri. Specifically, lead concentrations that were reported as part of monitoring for National Ambient Air Quality Standard (NAAQS) compliance indicate that lead is transported via the air pathway, and the measurements provide some indication of the sources responsible for the airborne lead. Nonattainment occurred in three counties—Iron, Dent, and Reynolds—in the Southeast Missouri Lead Mining District. One site includes the nearby secondary lead smelter (Boss, MO) that reported a 3-month running lead average of 0.17–0.18 $\mu g/m^3$ in April–August 2016.[2] Some monitoring sites in the facility fence line reported higher 3-month average values of 0.24–0.33 $\mu g/m^3$ from December 2016 through March 2017 (MO DNR 2017). An example of long-term variation in ambient lead concentrations in 2011–2016 is shown in Figure 3-6 and summarizes US Environmental Protection Agency (EPA) data. Although large decreases in total suspended particulate (TSP) lead concentrations in the Herculaneum and Doe Run areas were found after the smelter closed at the end of 2013, 588 samples showed that 24-hr lead concentrations exceeded the NAAQS threshold, 0.15 $\mu g/m^3$, in 2014–2016 (EPA 2017).

A fundamental characteristic of suspended particulate matter (PM) that affects transport is its size distribution, often expressed as a function of the particle's aerodynamic diameter.[3] The size distribution affects aerosol physics, chemistry, and transport, and environmental and health effects. This section first describes particle-size distributions and how they affect transport and then fugitive dust suspension and deposition, which are probably the most relevant to lead dispersal from mining-associated sites and other sources. The section concludes with an example of modeling of air transport.

Particle-Size Distribution and Implications for Transport

Airborne PM can range in diameter from a few nanometers to a few hundred micrometers, a span of 5 orders of magnitude, and follows a size distribution like that shown in Figure 3-7. Particles that have an aerodynamic diameter of greater than about 2.5 µm are called the coarse fraction, and particles that have an aerodynamic diameter of less than about 2.5 µm are called the fine fraction, which includes the nucleation, ultrafine, and accumulation fractions. Because mass scales with the cube of the diameter, PM *mass* concentrations are often dominated by the presence of large particles. Conversely, PM *number* concentrations are dominated by the smallest particles. The different size fractions have different sources, removal mechanisms, transport characteristics, atmospheric residence times, and deposition patterns in the environment and the respiratory tract.

Coarse particles (greater than 2.5 µm) are generated by mechanical action, such as crushing, grinding, material handling, wind erosion, or vehicular traffic (Watson et al. 2000). Large particles have high gravitational settling velocities and can deposit onto a surface within a few minutes to a few hours after suspension. For example, chat piles in Southeast Missouri consist of large particles, which are unlikely to become suspended and transported long distances by direct wind action except during extreme weather events, such as tornadoes. Particles that have diameters of a few micrometers, however, can remain suspended for several days and can therefore be transported long distances, even across continents or oceans. Atmospheric turbulence plays an important role in lofting suspended particles from near the ground to higher altitudes, thereby facilitating long-distance transport (Pye 1987).

Fine particles (less than 2.5 µm) are typically generated by condensation of hot vapors and by diffusion and coagulation of gases and other fine particles. Like smaller particles described above, fine particles can be transport-

[2]Nonattainment occurs when air quality is not in compliance with the NAAQS; that is, when the rolling 3-month average exceeds 0.15 $\mu g/m^3$.

[3]Aerodynamic diameter (d_a) varies inversely with the square root of the density. It is different from the geometric diameter (d_g); $d_g = d_a/\sqrt{\text{density}}$.

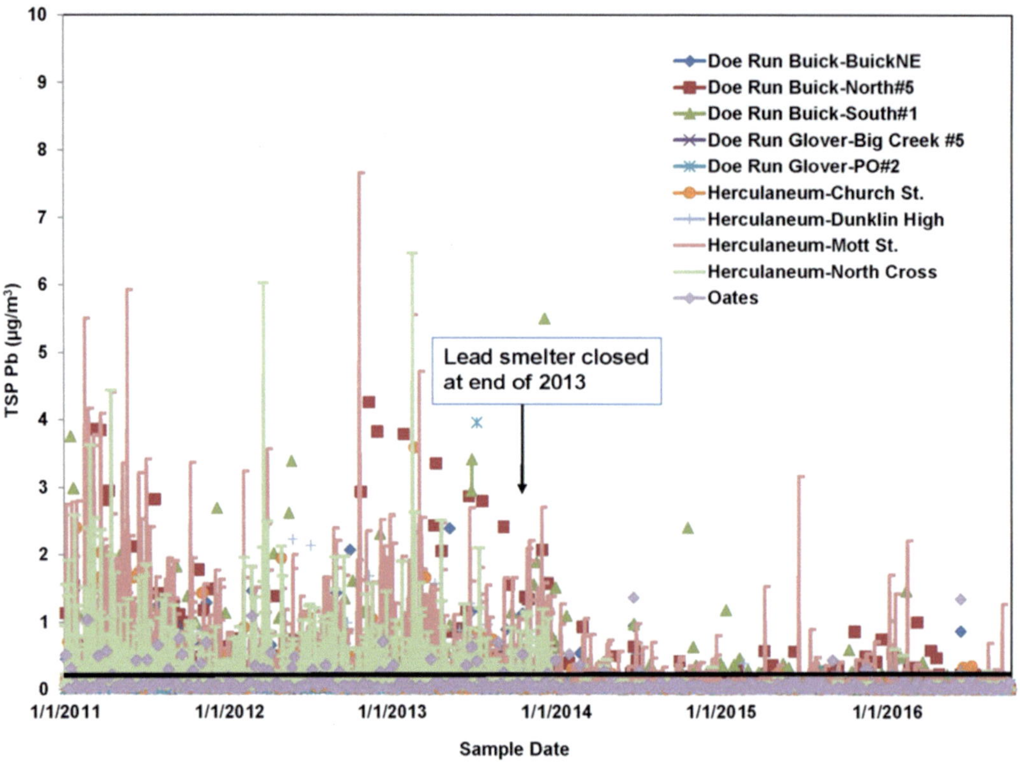

FIGURE 3-6 Time series of 24-hr average TSP lead concentrations at selected Missouri monitoring sites. The black line indicates the NAAQS threshold, 0.15 $\mu g/m^3$. Source: EPA (2017).

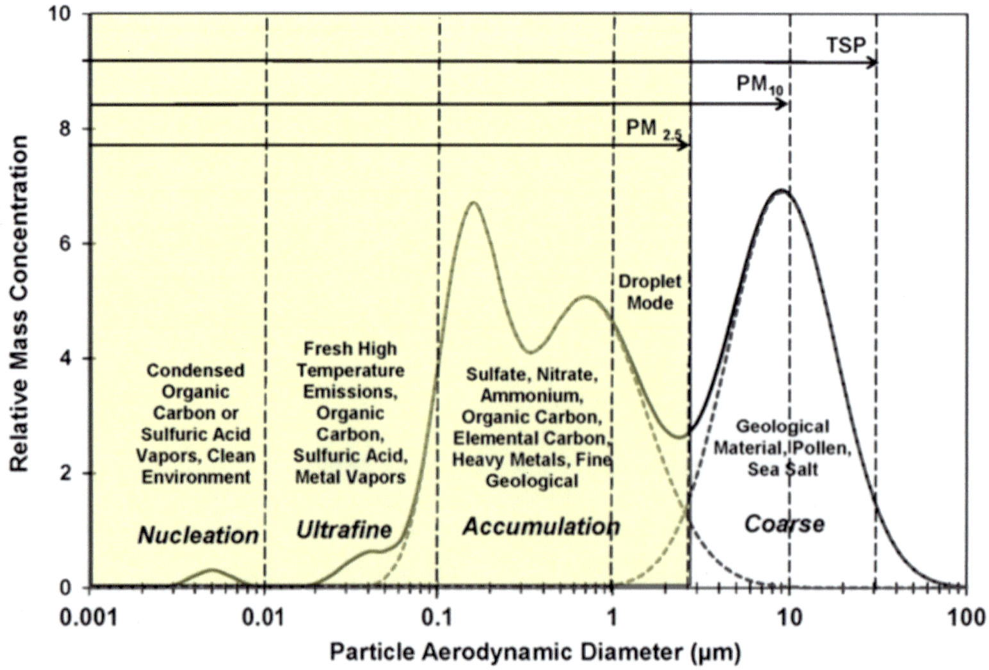

FIGURE 3-7 Illustration of different size fractions of airborne particles in a typical atmospheric particle-size distribution. Size-selective inlets with 50% collection efficiencies are used to determine aerodynamic diameters of $PM_{2.5}$ and PM_{10}. Sources and processes that affect each mode are indicated. Yellow-shaded area denotes fine-particle fraction (less than 2.5 µm). TSP refers to sizes up to about 30–50 µm. Source: Adapted from Chow (1995). Reprinted with permission; copyright 1995, *Journal of the Air & Waste Management Association*.

ed long distances. For example, lead-smelting activities in Southeast Missouri probably emitted lead-containing vapors that nucleated and condensed into fine particles (Moravec et al. 2015) or that condensed onto the surfaces of pre-existing fine particles; these particles were then transported long distances and deposited on the Earth's surface through both dry and wet deposition where they might still be present in soils.

Fugitive Dust Suspension and Mechanisms

Fugitive dust (Chow and Watson 1992; Kinsey and Cowherd 1992; Watson et al. 2000; Amato 2001; Cowherd 2001) consists of small particles that are suspended in the atmosphere after emission from such sources as mining pits, open fields and parking lots, paved and unpaved roads, agricultural fields, construction sites, dry lakes (playas), unenclosed storage piles, and material transfer systems. Visible dust plumes are often noticed over those sources when wind speeds are high or when vehicles are moving. Studies show that, on the average, fugitive dust contributes about 40–60% of PM_{10} and about 5–20% of $PM_{2.5}$ measured in the atmosphere (Watson and Chow 2000). However, fugitive dust contributions are often overestimated when published emission rates are used in dispersion models that simulate contributions to receptor concentrations (Watson et al. 2000) because the meteorologic, physical, and chemical factors on which fugitive dust emission rates are based can vary widely on local, regional, and national scales.[4] Resolving discrepancies between emission estimates and ambient measurements is an important consideration in designing, applying, and evaluating control strategies that are intended to reduce fugitive dust emissions.

As illustrated in Figure 3-8, dust suspension involves a complex three-step process: creep, saltation, and suspension. Creep is the rolling of the largest particles (about 500–1,000 μm) that are not suspended without extreme wind events; as the wind pushes them along, they can roll over and disaggregate other particles. Saltation is the movement of large particles (about 100–500 μm) that bounce a few centimeters above the surface and then fall back down, thereby bombarding the surface; upon impact, they dislodge smaller particles (less than 100 μm) from the surface, which can be entrained in the prevailing airflow. Creep, saltation, and suspension are inherently chaotic processes that are difficult to model (Shao 2008; Kok et al. 2012; Valance et al. 2015). Instead, semi-empirical models are used to estimate dust production from soils and stockpiles, such as chat piles and mine-tailings impoundments.

The suspension of fugitive dust particles depends on particle sizes, surface loadings, surface conditions, wind speeds, atmospheric and surface moisture, and the existence of dust-generating activities, such as vehicular traffic. Windblown dust occurs when the wind exceeds a threshold velocity. The minimum threshold wind velocity is often found to be about 5–10 m/s when extrapolated to 10 m above the surface (the standard height for reporting wind speed). However, the threshold can vary widely because of such factors as surface roughness, moisture content, vegetative cover, and particle size. Portable wind tunnels are increasingly used in the field to measure dust emissions and threshold wind velocities directly

[4]For further information on emission factors, see https://www.epa.gov/air-emissions-factors-and-quantification/basic-information-air-emissions-factors-and-quantification#About Emissions Factors.

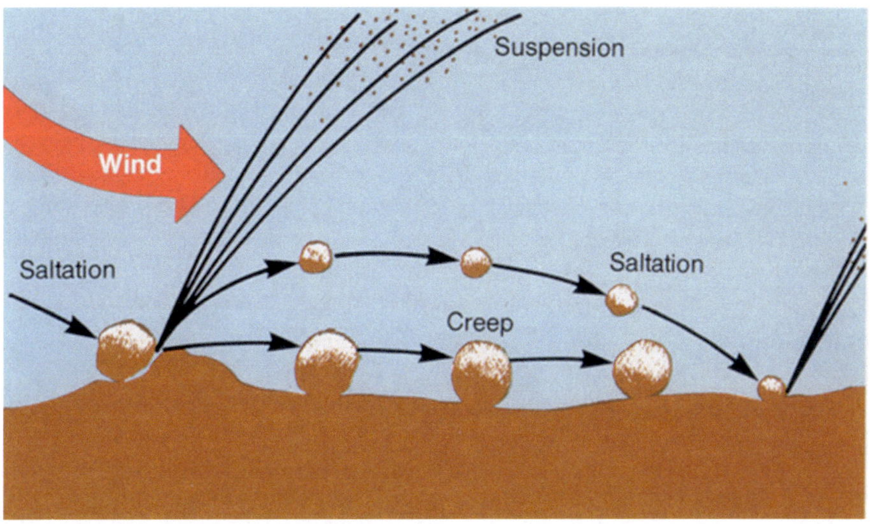

FIGURE 3-8 Example of windblown-dust suspension on a surface. Source: USDA (2015).

(Etyemezian et al. 2007). Figure 3-9 illustrates how the PM_{10} emissions can vary as a function of wind speed and surface conditions. Such emission measurements can be used in computational fluid-dynamics models to simulate atmospheric dispersion of dust (Stovern et al. 2014). Those models, for example, can be coupled to lead-isotope data to improve attribution of lead to various sources (Stovern et al. 2016).

Mechanical suspension processes are dominated by moving vehicles, wherein the contact of a tire with the surface of a roadway grinds the dust to smaller sizes and injects it from the road surface into the atmosphere. Vehicle wakes that create turbulence are also sources of dust suspension, although they can also be classified as wind-induced. Other mechanical suspension methods involve transfer of bulk materials through uncovered conveyors, bulldozers, shovels, and open trucks.

Modeling Air Transport: An Example from Southeast Missouri

An example from Southeast Missouri illustrates the potential for air transport of lead-containing particles from chat piles and tailings. Abbott (1999) modeled the erosion and transport of airborne PM from chat piles and tailings impoundments at Bonne Terre, Desloge, Leadwood, National, Elvins, and Federal in Southeast Missouri. The model was evaluated using data from air-quality monitors—presumably TSP monitors, but that is not stated—at four locations and downwind soil sampling. As expected, predicted transport from coarse chat piles (0.5–1.0 km) was less than transport from the fine tailings (3–5 km). The highest deposition rates were predicted for areas east and southeast of the sources, which would be consistent with the prevailing wind direction. The maximum annual average calculated lead deposition rate 1 km downwind from tailings impoundments was about 1–30 $\mu g/m^2$-hr. Although model predictions and atmospheric observations varied from site to site, the predicted airborne lead at the Desloge site was reasonably close to observations with respect to both annual average (0.1 $\mu g/m^3$) and maximum 24-h average (1 $\mu g/m^3$) concentrations. Soil sampling showed the expected rapid decrease in lead concentration with distance that reached more typical concentrations at about 2 km downwind from the Bonne Terre sources, which were mainly tailings and a small chat pile.

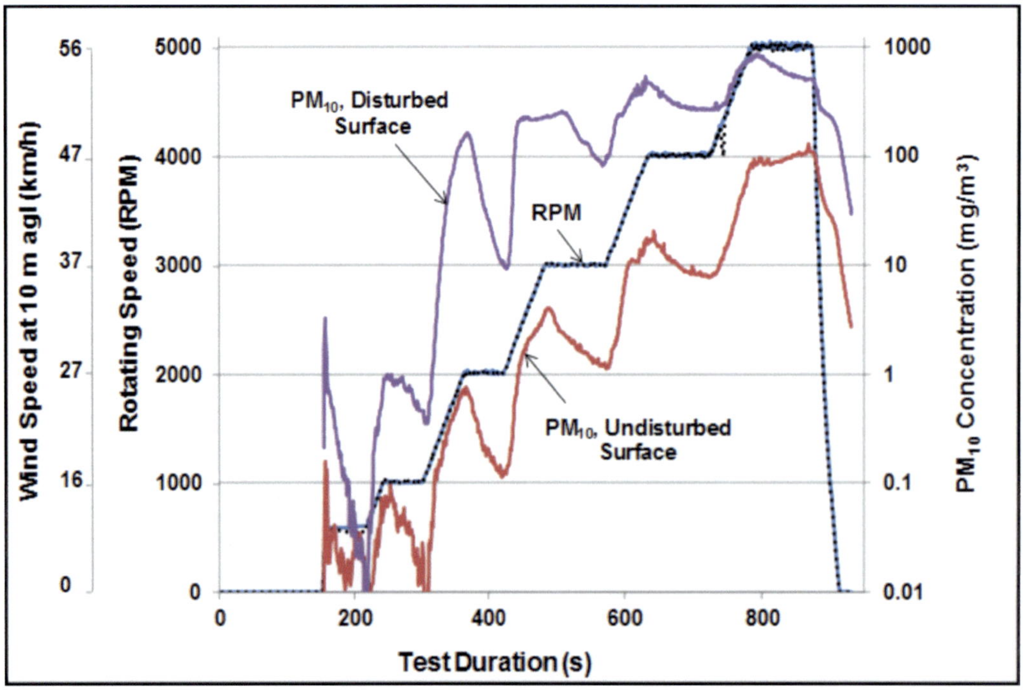

FIGURE 3-9 Comparison of PM_{10} suspended at different wind speeds from mechanically disturbed surfaces (such as a broken surface crust) and undisturbed surfaces of a coke storage pile; data collected using a portable wind tunnel (Etyemezian et al. 2007). Emissions increase with wind speed for both disturbed and undisturbed surfaces, but the emissions are more than an order of magnitude lower for the undisturbed surface. Source: Adapted from Watson et al. (2014). Reprinted with permission; copyright 2014, Desert Research Institute.

A decade later, the Missouri Department of Natural Resources reported that airborne lead concentrations monitored near a chat pile during earth-moving operations could rise to 0.3 μg/m^3 (24-hr average) despite the coarse particle diameter of chat (MO DNR 2009). High lead observations from TSP monitors were also reported at St. Joe State Park (up to 0.9 μg/m^3) and from personal monitors (up to 6 μg/m^3, time-weighted average) worn by motorcycle riders in the vicinity of the disturbed surface of the chat pile (MO DNR 2009). Such observations near chat sources show that in spite of the very large particle diameter, chat can be an important lead source where anthropogenic mechanical disturbance occurs. That observation is important because chat was spread on roads, where the coarse particles can be ground up by traffic, and then suspended by the passage of vehicles or wind erosion. The grinding and resuspension of chat could be an important lead source; it has been reported that some road dust in Southeast Missouri contains higher lead concentrations than are measured elsewhere in the region (Witt et al. 2013, 2014, 2016).

The Abbott (1999) study used meteorologic observations from the St. Louis airport 80–100 km away; wind was predominantly toward the east and southeast, but these wind directions probably do not reflect local conditions, where the topography might channel the flow in a more northerly or southerly direction. The study demonstrates the importance of making local meteorologic observations for source dispersion modeling, as discussed further in Chapter 4.

INTENTIONAL TRANSPORT

Intentional transport includes all the human activities that move materials within and off a site. For example, intentional transport at mining sites in the Southeast Missouri Lead Mining District would include the mechanical, truck, or rail transport of lead ores, lead concentrates, or lead-containing mine wastes (waste rock, tailings, and slags) for further processing, disposal, or beneficial use, such as use of chat as soil amendments on agricultural fields and use of tailings as railroad-bed material. McHenry (2006) compiled a history of the mining companies and chat dumps of the Missouri Lead Belt in St. Francois County and reported considerable anecdotal information about transport of mine wastes, particularly chat, for beneficial uses. Before 1922, chat was offered "free for the asking and many thousands of tons were hauled away by local citizens, city, county and state government, and other industries in the area which had uses for it." As noted in Chapter 2, one primary use was as an agricultural lime; in 1922, chat began to be sold to limestone dealers for use as a soil amendment. Chat sales continued until 1972, when one chat pile that originally contained 3,522,000 tons was entirely gone (Karsch 1973). Other uses noted by McHenry (2006) were as a component of blacktop, concrete, fill material in building construction and road grades, sandbox sand, and grit for deicing streets and sidewalks.

The intentional transport of mine wastes for beneficial use is of concern as a potential vector for public-health and environmental problems. The Comprehensive Environmental Response, Compensation and Liability Act (CERCLA) remedial investigations (RIs) for mining sites and other industrial sites that have generated large quantities of potentially valuable byproducts do not systematically consider intentional transport mechanisms. Because intentional transport is a business activity, the process of identifying and quantifying intentional transport processes involves locating and auditing business records, such as transport manifests, invoices and receipts, inventory records, and informal communications (oral and written). The systematic consideration of intentional transport mechanisms then falls under the purview of accountants, economists, and historians, who are rarely engaged in such activity for CERCLA RIs, perhaps because of an absence of EPA guidance that articulates the need for and role of qualified practitioners in these professions in systematically considering intentional transport mechanisms.[5]

One approach for encouraging systematic consideration of intentional transport would be to eliminate the use of the term *migration* to describe the transport processes to be considered because its use implies that contaminant transport is governed only by air and water dispersion processes. Furthermore, a recommendation to consider intentional transport mechanisms could be added to guidance documents.

REFERENCES

Abbott, M.L. 1999. Air Dispersion Modeling Of Mine Waste in the Southeast Missouri Old Lead Belt. INEEL/EXT-99-00235. Idaho National Engineering and Environmental Laboratory and Environmental Laboratory Integrated Earth Sciences Department, Idaho Falls, ID. October 1999 [online]. Available: https://inldigitallibrary.inl.gov/sites/sti/sti/3156874.pdf [accessed July 11, 2017].

Amato, J.A. 2001. Dust: A History of the Small and the Invisible. Berkeley, CA: University of California Press.

Bain, D.J., M.B. Green, J.L. Campbell, J.F. Chamblee, S. Chaoka, J.M. Fraterrigo, S.S. Kaushal, S.L. Martin, T.E. Jordan, A.J. Parolari, W.V. Sobczak, D.E. Weller, W.M. Wolheim, E.R. Boose, J.M. Duncan, G.M. Gettel, B.R.

[5]The committee notes that archeologists and historians are routinely enlisted at CERCLA sites to locate and document historical artifacts, as required to comply with the National Historic Preservation Act, but social scientists are not routinely enlisted to consider intentional transport mechanisms and their contribution to contaminant distribution systematically.

Hall, P. Kumar, J.R. Thompson, J.M. Vose, E.M. Elliott, and D.S. Leigh. 2012. Legacy effects in material flux: Structural catchment changes predate long-term studies. BioScience 62(6):575-584.

Bigham, J.M., and D.K. Nordstrom. 2000. Iron and aluminum hydroxysulfates from acid sulfate waters. Rev. Mineral. Geochem. 40(1):351-403.

Brewer, P.A., and M.P. Taylor. 1997. The spatial distribution of heavy metal contaminated sediment across terraced floodplains. Catena 30(2-3):229-249.

Callender, E. 2005. Heavy metals in the environment—Historical trends. Pp. 67-105 in Environmental Geochemistry, Treatise on Geochemistry Vol. 9, B. Sherwood Lollar, ed. Amsterdam: Elsevier.

Cao, X., L.Q. Ma, M. Chen, D.W. Hardison, Jr., and W.G. Harris. 2003a. Weathering of lead bullets and their environmental effects at outdoor shooting ranges. J. Environ. Qual. 32:526-534.

Cao, R.X., L.Q. Ma, M. Chen, S.P. Singh, and W.G. Harris. 2003b. Phosphate-induced metal immobilization in a contaminated site. Environ. Pollut. 122(1):19-28.

Chow, J.C. 1995. Measurement methods to determine compliance with ambient air quality standards for suspended particles. J. Air Waste Manage. Assoc. 45(5):320-382.

Chow, J.C., and J.G. Watson. 1992. Fugitive emissions add to air pollution. Environ. Protect. 3:26-31.

Cowherd, C. 2001. Fugitive dust emissions. Pp. 845-857 in Aerosol Measurement: Principles, Techniques, and Applications, 2nd Ed., P. Baron, and K. Willeke, eds. New York: John Wiley & Sons.

Dutrizac, J.E., and J.L. Jambor. 2000. Jarosites and their application in hydrometallurgy. Rev. Mineral. Geochem. 40(1):405-452.

EPA (U.S. Environmental Protection Agency). 1996. Method 1669. Sampling Ambient Water for Trace Metals at EPA Water Quality Criteria Levels. July 1996 [online]. Available: https://www3.epa.gov/caddis/pdf/Metals_Sampling_EPA_method_1669.pdf [accessed August 4, 2017].

EPA. 2017. Interactive Map of Air Quality Monitors [online]. Available: https://www.epa.gov/outdoor-air-quality-data/interactive-map-air-quality-monitors [accessed June 22, 2017].

Etyemezian, V.R., G. Nikolich, S. Ahonen, M.L. Pitchford, M. Sweeney, R. Purcell, J.A. Gillies, and H.D. Kuhns. 2007. The Portable In Situ Wind Erosion Laboratory (PI-SWERL): A new method to measure PM10 potential for windblown dust properties and potential for emissions. Atmos. Environ. 41(18): 3789-3796.

Gilbert, G.K. 1917. Hydraulic-Mining Debris in the Sierra Nevada. U.S. Geological Survey Professional Paper 105. Washington, DC: U.S. Government Printing Office [online]. Available: https://pubs.usgs.gov/pp/0105/report.pdf [accessed June 22, 2017].

Graf, W.L., S.L. Clark, M.T. Kammerer, T. Lehman, K. Randall, R. Tempe, and A. Schroeder. 1991. Geomorphology of heavy metals in the sediments of Queen Creek, Arizona, USA. Catena 18(6):567-582.

Guilbert, J.M., and C.F. Park, Jr. 1986. The Geology of Ore Deposits. New York: WH Freeman.

Gutierrez, M., K. Mickus, and L.M. Camacho. 2016. Abandoned Pb-Zn mining wastes and their mobility as proxy to toxicity: A review. Sci. Total Environ. 565:392-400.

Hardison Jr., D.W., L.Q. Ma, T. Luongo, and W.G. Harris. 2004. Lead contamination in shooting range soils from abrasion of lead bullets and subsequent weathering. Sci. Total Environ. 328(1-3):175-183.

Hayes, S.M., S.A. White, T.L. Thompson, R.M. Maier, and J. Chorover. 2009. Changes in lead and zinc lability during weathering-induced acidification of desert mine tailings: Coupling chemical and micro-scale analyses. Appl. Geochem. 42(12):2234-2245.

Hem, J.D. 1976. Inorganic chemistry of lead in water. Pp. 5-11 in Lead in the Environment, T.G. Lovering, ed. Geological Survey Professional Paper 957. Washington, DC: U.S. Geological Survey [online]. Available: https://pubs.usgs.gov/pp/0957/report.pdf [accessed June 22, 2017].

Hillier, S. 2001.Particulate composition and origin of suspended sediment in the R. Don, Aberdeenshire, UK. Sci. Total Environ. 265(1-3):281-293.

Horowitz, A. 1991. A Primer on Sediment-Trace Element Chemistry, 2nd Ed. Chelsea, MI: Lewis.

Hudson-Edwards, K.A., M.G. Macklin, C.D. Curtis, and D.J. Vaughan. 1996. Processes of formation and distribution of Pb-, Zn-, Cd-, and Cu-bearing minerals in the Tyne Basin, northeast England: Implications for metal-contaminated river systems. Environ. Sci. Technol. 30(1):72-80.

Hudson-Edwards, K.A., M.G. Macklin, J.R. Miller, and P.J. Lechler. 2001. Sources, distribution and storage of heavy metals in the Rio Pilcomayo, Bolivia. J. Geochem. Explor. 72: 229-250.

Jambor, J.L., D.K. Nordstrom, and C.N. Alpers. 2000. Metal-sulfate salts from sulfide mineral oxidation. Rev. Mineral. Geochem. 40(1):305-350.

James, L.A. 1989. Sustained storage and transport of hydraulic gold mining sediment in the Bear River, California. Ann. Assoc. Am. Geogr. 79(4):570-592.

Jørgensen, S.S., and M. Willems. 1987. The fate of lead in soils: The transformation of lead pellets in shooting-range soils. Ambio 16(1):11-15.

Karsch, A.1973. Chat dumps of St. Francois County, Missouri, lead mining capital of the World: 1864-1972. The Farmington News, January. P. xv in Chat Dumps of the Missouri Lead Belt, St. Francois County, with an Illustrated History of the Lead Companies that Built Them, 2006. Park Hills, MO: Missouri Department of Natural Resources, Missouri Mines State Historic Site.

Kinsey, J.S., and C. Cowherd. 1992. Fugitive emissions. Pp. 133-146 in Air Pollution Engineering Manual, A.J. Buonicore, and W.T. Davis, eds. New York: Van Nostrand Reinhold.

Knighton, D. 1991. Channel bed adjustment along mine-affected rivers of northeast Tasmania. Geomorphology, 4:205-219.

Knox, J.C. 1987. Historical valley floor sedimentation in the Upper Mississippi valley. Ann. Assoc. Am. Geogr. 77(2):224-244.

Knox, J. C. 1989. Valley floor sedimentation in the Upper Mississippi Valley: Reply. Ann. Assoc. Am. Geogr. 79(4):601-608.

Kok, J.F., E.J.R. Parteli, T.I. Michaels, and D.B. Karam. 2012. The physics of wind-blown sand and dust. Rep. Prog. Phys. 75:106901 [online]. Available: https://arxiv.org/ftp/arxiv/papers/1201/1201.4353.pdf [accessed June 23, 2017].

Leigh, D.S. 2010. Morphology and channel evolution of small streams in the southern Blue Ridge Mountains of Western North Carolina. Southeast. Geogr. 50(4):397-421.

Lewin, J., and M.G. Macklin. 1987. Metal mining and floodplain sedimentation in Britian. Pp. 1009-1027 in International Geomorphology, V. Gardiner, ed. Hoboken, NJ: John Wiley and Sons.

Lin, Z., B. Comet, U. Qvarfort, and R. Herbert. 1995. The chemical and mineralogical behavior of Pb in shooting range soils from central Sweden. Environ. Pollut. 89(3):303-309.

Lovering, T.G. 1976. Lead in the environment – Summary. Pp. 1-4. in Lead in the Environment, T.G. Lovering, ed. Geological Survey Professional Paper 957. Washington, DC: U.S. Geological Survey [online]. Available: https://pubs.usgs.gov/pp/0957/report.pdf [accessed June 22, 2017].

Mackin, M.G. 1996. Fluxes and storage of sediment-associated heavy metals in floodplain systems: Assessment and river basin management issues at a time of rapid environmental change. Pp. 441-460 in Floodplain Processes, M.G. Anderson, D.E. Walling, and P.D. Bates, eds. Hoboken, NJ: John Wiley and Sons.

Macklin, M.G., and R.B. Dowsett. 1989. The chemical and physical speciation of trace metals in fine grained overbank flood sediments in the Tyne Basin, north-east England. Catena 16(2):135-151.

Macklin, M.G., and J. Lewin. 1989. Sediment transfer and transformation of an alluvial valley floor: The river South Tyne, Northumbria, U.K. Earth. Surf. Proc. Land. 14(3):233-246.

Macklin, M.G., K.A. Hudson-Edwards, H.E. Jamieson, P. Brewer, T.J. Coulthard, A.J. Howard, and V.H. Renenda. 1999. Physical stability and rehabilitation of sustainable aquatic and riparian ecosystems in the Rio Guadiamar, Spain, following the Aznalcóllar mine tailings dam failure. Pp. 271-278 in Mine, Water and Environment: Proceedings of the International Mine Water Association Congress, Seville, Spain [online]. Available: http://www.mwen.info/docs/imwa_1999/IMWA1999_Macklin_271.pdf [accessed June 23, 2017].

Macklin, M.G., P.A. Brewer, K.A. Hudson-Edwards, G. Bird, T.J. Coulthard, I.A. Dennis, P.J. Lechler, J.R. Miller, and J.N. Turner. 2006. A geomorphological-geochemical approach to river basin management in mining-affected rivers. Geomorphology 79:423-447.

Marron, D.C. 1987. Floodplain storage of metal-contaminated sediments downstream of a gold mine at Lead, South Dakota. Pp. 193-209 in Chemical Quality of Water and the Hydrologic Cycle, R.C. Averett, and D.M. McKnight, eds. Chelsea, MI: Lewis.

Marzouk, E.R., S.R. Chenery, and S.D. Young. 2013. Predicting the solubility and lability of Zn, Cd, and Pb in soils from a minespoil-contaminated catchment by stable isotopic exchange. Geochim. Cosmochim. Acta 123:1-16.

Masters, S., G.J. Welter, and M. Edwards. 2016. Seasonal variations in lead release to portable water. Environ. Sci. Technol. 50(10):5269-5277.

McHenry, R.E. 2006. Chat Dumps of The Missouri Lead Belt, St. Francois County, With an Illustrated History of the Lead Companies that Built Them. Park Hills, MO: Missouri Department of Natural Resources, Missouri Mines State Historic Site.

Miller, J.R., and S.M. Orbock Miller. 2007. Contaminated Rivers: A Geomorphological-Geochemical Approach to Site Assessment and Remediation. Berlin: Springer.

Miller, J.R., and L.F. Villarroel. 2011. Bolivia: Mining, river contamination and human health. Pp. 421-441in Encyclopedia of Environmental Health, J. Nriagu, ed. Amsterdam: Elsevier.

Miller, J.R., P.J. Lechler, and M. Desilets. 1998. The role of geomorphic processing in the transport and fate of mercury in the Carson River basin, west-central Nevada. Environ. Geol. 33(4):249-262.

Miller, J.R., D. Germanoski, L.F. Villarroel, and P. Lechler. 2009. Spatial and temporal variations in the transport and storage of trace metal contaminants in the Upper Río Pilcomayo, Southern Bolivia. Int. J. Environ. Health 3(4):334-361.

Miller, J.R., M. Lord, L. Villarroel, D. Germanoski, and J. Chambers. 2012. Structural organization of process zones in upland watersheds of central Nevada and its influence on basin connectivity, dynamics, and wet meadow complexes. Geomorphology 139-140:284-402.

MO DNR (Missouri Department of Natural Resources). 2009. Air Pollution Control Program: Missouri Lead Monitoring Network Plan, October 2009 [online]. Available: https://dnr.mo.gov/env/apcp/monitoring/moleadplan-09.pdf [accessed June 23, 2017].

MO DNR. 2017. Air Quality Analysis for Lead [online]. Available: https://dnr.mo.gov/env/apcp/docs/leadmonitordata.pdf [accessed July 12, 2017].

Moore, J.N., E.J. Brook, and C. Johns. 1989. Grain size partitioning of metals in contaminated, coarse-grained river floodplain sediments: Clark Fork River, Montana, USA. Environ. Geol. Water S. 14(2):107-115.

Moravec, P., J. Smolik, J. Ondracek, P. Vodicka, and R. Fajgar. 2015. Lead and/or lead oxide nanoparticle generation for inhalation experiments. Aerosol Sci. Technol. 49(8):655-665.

Nicholas, A.P., P.J. Ashworth, M.J. Kirkby, M.G. Macklin, and T. Murray. 1995. Sediment slugs: Large-scale fluctuations in fluvial sediment transport rates and storage volumes. Prog. Phys. Geogr. 19(4):500-519.

Niemitz, J., C. Haynes, and G. Lasher. 2013. Legacy sediments and historic land use: Chemostratigraphic evidence for excess nutrient and heavy metal sources and remobilization. Geology 41(1):47-50.

Nieto, J.M., A.M. Sarmiento, M. Olías, C.R. Canovas, I. Riba, J. Kalman, and T.A. Delvalls. 2007. Acid mine drainage pollution in the Tinto and Odiel rivers (Iberian Pyrite Belt, SW Spain) and bioavailability of the transported metals to the Huelva Estuary. Environ. Int. 33(4):445-455.

Nordstrom, D.K., D.W. Blowes, and C.J. Ptacek. 2015. Hydrogeochemistry and microbiology of mine drainage: An update. Appl. Geochem. 57:3-16.

Orbock Miller, S.O., D.F. Ritter, R.C. Kochel, and J.R. Miller. 1993. Fluvial responses to land-use changes and climatic variations within the drury creek watershed, southern Illinois. Geomorphology 6(4):309-329.

Pavlowsky, R.T., M.R. Owen, and D.J. Martin. 2010. Big River Mining Sediment Assessment Project: Distribution, Geochemistry, and Storage of Mining Sediment in Channel and Floodplain Deposits of the Big River System in St. Francois, Washington, and Jefferson Counties, Missouri, Final Report. OEWRI EDR-10-002. Missouri State University, Ozarks Environmental and Water Resources Institute, Springfield, MO. June 18, 2010.

Pieper, K.J., M. Tang, and M.A. Edwards. 2017. Flint water crisis caused by interrupted corrosion control: Investigating "ground zero" home. Environ. Sci. Technol. 51(4):2007-2014.

Powell, K.J., P.L. Brown, R.H. Byrne, T. Gajda, G. Hefter, A.K. Leuz, S. Sjöberg and H. Wanner. 2009. Chemical speciation of environmentally significant metals with inorganic ligands, Part 3. The Pb2+ + OH-, Cl-, CO32-, SO42-, and PO43-–systems. IUPAC Technical Report. Pure Appl. Chem. 81(12):2425-2476.

Prabhakar, G., A. Sorooshian, E. Toffol, A.F. Arellano, and E.A. Betterton. 2014. Spatiotemporal distribution of airborne particulate metals and metalloids in a populated arid region. Atmos. Environ. 92:339-347.

Price, K., and D. Leigh. 2006. Morphological and sedimentological responses of streams to human impact in the southern Blue Ridge Mountains, USA. Geomorphology 78(1-2):142-160.

Pye, K. 1987. Aeolian Dust and Dust Deposits. London, UK: Academic Press.

Richards, R.P. 2001. Estimation of Pollutant Loads in Rivers and Streams: A Guidance Document for NPS Programs. Report to the Environmental Protection Agency, Region VIII, Grant No. X998397-01-0 [online]. Available: http://www.ccamp.us/ccamp_regional_report/_Documents/PeteLoad_Est1.pdf [accessed June 26, 2017].

Rooney, C.P., R.G. McLaren, and L.M. Condron. 2007. Control of lead solubility in soil contaminated with lead shot: Effect of soil pH. Environ. Pollut. 149(2):149-157.

Root, R.A., S.M. Hayes, C.M. Hammond, R.M. Maier, and J. Chorover. 2015. Toxic metal(loid) speciation during weathering of iron sulfide mine tailings under semi-arid climate. Appl. Geochem. 62:131-149.

Rouff, A.A., E.J. Elzinga, R.J. Reeder, and N.S. Fisher. 2005. The influence of pH on the kinetics, reversibility and mechanisms of Pb(II) sorption at the calcite-water interface. Geochim. Cosmochim. Acta 69(22):5173-5186.

Rozan, T.F., G.W. Luther, D. Ridge, and S. Robinson. 2003. Determination of Pb complexation in oxic and sulfidic waters using pseudovoltammetry. Environ. Sci. Technol. 37(17):3845-3852.

Schaider, L.A., D.B. Senn, E.R. Estes, D.J. Brabander, and J.P. Shine. 2014. Sources and fates of heavy metals in a mining-impacted stream: Temporal variability and the role of iron oxides. Sci. Total Environ. 490:456-466.

Serrano, S., P.A. O'Day, D. Vlassopoulos, M.T. Garcia-Gonzalez, and F. Garrido. 2009. A surface complexation and ion exchange model of Pb and Cd competitive sorption on natural soils. Geochim. Cosmochim. Acta 73:543-558.

Serrano, S., D. Vlassopoulos, F. Garrido, and P.A. O'Day. 2013. A combined site-specific metals sorption and transport model for intact soil columns. Vadose Zone J. 12(4), doi:10.2136/vzj2012.0152.

Shao, Y.P. 2008. Physics and Modelling of Wind Erosion. New York: Springer.

Slingerland, R., and N.D. Smith. 1986. Occurrence and formation of water-laid placers. Annu. Rev. Earth Planet Sci. 14:113-147.

Stockdale, A., E. Tipping, and S. Lofts. 2015. Dissolved trace metal speciation in estuarine and coastal waters: Comparison of WHAM/Model VII predictions with analytical results. Environ. Sci. Technol. 34(1):53-63.

Stovern, M., O. Felix, J. Csavina, K.P. Rine, M.R. Russell, R.M. Jones, M. King, E.A. Betterton, and A.E. Sáez. 2014. Simulation of windblown dust transport from a mine tailings impoundment using a computational fluid dynamics model. Aeolian Res. 14:75-83.

Stovern, M., H. Guzmán, K.P. Rine, O. Felix, M. King, W.P. Ela, E.A. Betterton, and A.E. Sáez. 2016. Windblown dust deposition forecasting and spread of contamination around mine tailings. Atmosphere 7(2):16.

Stumm, W. 1992. Chemistry of the Solid-Water Interface: Processes at the Mineral-Water and Particle-Water Interface in Natural Systems. New York: Wiley [online]. Available: http://www.jlakes.org/config/hpkx/news_category/2015-12-11/ChemistryOfTheSolidWaterInterface_Stumm1992.pdf [accessed August 4, 2017].

Swistock, B.R., W.E. Sharpe, and P.D. Robillard. 1993. A survey of lead, nitrate and radon contamination of private individual water systems in Pennsylvania. J. Environ. Health 55:6-12.

Town, R.M., and M. Filella. 2000a. A comprehensive systematic compilation of complexation parameters reported for trace metals in natural waters. Aquat. Sci. 62(3):252-295.

Town, R.M., and M. Filella. 2000b. Dispelling the myths: Is the existence of L1 and L2 ligands necessary to explain metal ion speciation in natural waters? Limnol. Oceanogr. 45(6):1341-1357.

Trimble, S.W. 1974. Man-induced Soil Erosion in the Southern Piedmont 1700-1970. Ankeny, IA: Soil Conservation Society of America. 180pp.

USDA (U.S. Department of Agriculture). 2015. Wind erosion simulation models. U.S. Department of Agriculture, Wind Erosion Research, Fort Collins, CO [online]. Available: https://infosys.ars.usda.gov/WindErosion/simmodels/simmodels.shtml [accessed June 26, 2017].

Valance, A., K.R. Rasmussen, A.O. El Moctar, and P. Dupont. 2015. The physics of Aeolian sand transport. C. R. Phys. 16(1):105-117.

Vega, F.A., and L.P. Weng. 2013. Speciation of heavy metals in River Rhine. Water Res. 47(1):363-372.

Villarroel, L.F., J.R. Miller, P.J. Lechler, and D. Germanoski. 2006. Lead, zinc, and antimony contamination of the Rio Chilco-Rio Tupiza drainage system, southern Bolivia. Environ. Geol. 51(2):283-299.

Watson, J.G., and J.C. Chow. 2000. Reconciling Urban Fugitive Dust Emissions Inventory and Ambient Source Contribution Estimates: Summary of Current Knowledge and Needed Research. Desert Research Institute Document No. 6110.4F. Prepared for EPA, Research Triangle Park, NC. May 2000 [online]. Available: http://www.epa.gov/ttn/chief/efdocs/fugitivedust.pdf.

Watson, J.G., J.C. Chow, and T.G. Pace. 2000. Fugitive dust emissions. Pp. 117-135 in Air Pollution Engineering Manual 2nd Ed., W.T. Davis, ed. New York: John Wiley & Sons.

Watson, J.G., J.C. Chow, X. Wang, S.D. Kohl, and L.N.R. Yatavelli. 2014. Windblown Fugitive Dust Characterization in the Athabasca Oil Sands Region. Prepared by Desert Research Institute, Reno, NV, for Wood Buffalo Environmental Association, Ft. McMurray, AB, Canada.

Witt, E.C., D.J. Wronkiewicz, R.T. Pavlowsky, and H.L.Shi. 2013. Trace metals in fugitive dust from unsurfaced roads in the Viburnum Trend resource mining District of Missouri-Implementation of a direct-suspension sampling methodology. Chemosphere 92:1506-1512.

Witt, E.C., H.L. Shi, D.J. Wronkiewicz, and R.T. Pavlowsky. 2014. Phase partitioning and bioaccessibility of Pb in suspended dust from unsurfaced roads in Missouri-A potential tool for determining mitigation response. Atmos. Environ. 88:90-98.

Witt, E.C., M.J. Pribil, J.P. Hogan, and D.J. Wronkiewicz. 2016. Isotopically constrained lead sources in fugitive dust from unsurfaced roads in the southeast Missouri mining district. Environ. Pollut. 216:450-459.

Xie, Y., and D.E. Giammar. 2011. Effects of flow and water chemistry on lead release rates from pipe scales. Water Res. 45(19):6525-6534.

Yang, J.Y., X.E. Yang, Z.L. He, I.Q. Li, J.L. Shentu, and P.J. Stoffella. 2006. Effect of pH, organic acids and inorganic ions on lead desorption from soils. Environ. Pollut. 143(1):9-15.

4

Investigative Strategies for Lead-Source Attribution

During the last several decades, a wide array of analytic techniques have been used, with varied degrees of success, to determine the relative contributions of lead from natural and anthropogenic sources in soils, sediments, water, and air. The US Environmental Protection Agency has used several methods in its investigations, including using elemental concentrations and ratios, isotopic analysis, and other soil-metal characterizations (Burgess 2016). Most techniques rely on a fingerprinting approach that attempts to identify unique physical or chemical characteristics that can distinguish lead or lead-associated materials between suspected sources (Box 4-1). Once the sources have been assessed, the distinguishing source characteristics can be compared with those of samples of soil, sediment, water, and air to assess the origin of the lead. Given the variety of analytic methods that are available, integration of the most appropriate techniques is challenging (Gallego et al. 2016).

In this chapter, the committee provides a series of investigative strategies that are stratified by media type. The strategies are intended to provide a logical framework for determining the source of lead while minimizing the time, effort, and financial resources required to do so. They are not intended to replace procedures associated with the remedial investigation and feasibility study but are meant to serve as a general guide to lead-source attribution. Given the highly variable nature of lead contamination near and adjacent to Superfund sites, not all the methods that are listed in the strategies will be applicable to all sites. In some cases, additional methods not cited in the strategies might be useful. Furthermore, the strategies might require modification to fit the specific characteristics of an individual site or operable unit.

The suggested strategies are based on a multiple-lines-of-evidence approach, in which data are systematically collected, analyzed, and interpreted until there is sufficient confidence that a method (or combination of methods) has resulted in a sound understanding of the sources of lead. The methods inherent in the strategies represent tested techniques that generate results that are widely accepted by the scientific community (unless otherwise stated). Preference was also given to techniques that generate quantifiable degrees of uncertainty.

SOIL

The committee's investigative strategy for soil is shown in Figure 4-1 and provides a guiding framework for deducing lead sources in soil. In this section, the committee focuses on soils derived from underlying rock strata as opposed to materials that are transported by river, gravity, or wind. The section that follows focuses on river sediments and sites within active floodplains where alluvial deposition is occurring regularly or has occurred over the duration in question for lead deposition.

The committee provides a possible sequence of analyses for soil based on typical successful outcomes, but the sequence might be adjusted on the basis of site-specific conditions, availability of instrumentation, and resource (cost and time) considerations. Before investigation of the source of lead, total soil concentrations need to be measured; sites (or samples) exceeding a regulatory threshold can then be examined for the lead source.

Site Characterization and History

Before a detailed chemical or mineralogic analysis is conducted, a basic site characterization and history should be developed that leads to a site conceptual model that can guide investigations toward likely lead sources; potential contributors of lead noted in Chapter 2 should be specifically identified. The landform of the soil or yard should be noted in terms of its current geology and likely geologic past. The US Department of Agriculture soil survey should be consulted for information on soils and landforms of the area. A history of land use at the site should also be constructed from public records and on-site observations to address appropriate questions, such as, Are there or have there been buildings constructed before the 1980s on the site that could have contributed lead-based paint to the

BOX 4-1 Lead Attribution Using Physical and Geochemical Fingerprinting Techniques

The attribution of lead in soils, sediments, water, and air commonly relies on the application of physical and geochemical "fingerprinting" methods. Although the approach varies widely with the media under investigation, the general assumption is that the physical or chemical composition of a sample is a function of the characteristics or composition of the lead sources and the relative amount of each source that contributes to it. This situation is referred to as an inverse (or unmixing) problem (Weltje 2012). Solution of the problem generally involves a four-step process (Zhang et al. 2012): (1) delineation and characterization of the lead sources, (2) identification of the physical or chemical properties (the fingerprint) that can distinguish the identified lead sources, (3) collection and characterization of samples from the media or site under consideration, and (4) determination of the relative contribution of lead by each source on the basis of some form of mathematical model. The latter step also includes an assessment of uncertainty associated with the modeling results.

Ideally, the fingerprint represents a unique property or set of properties that separate one source from all other sources. However, unlike a human fingerprint, the fingerprint of a lead source is, in many cases, imperfect. Because of inhomogeneity in the physical and chemical composition of the lead sources, the value or characteristics of the fingerprinting properties might be statistically distinct but overlap. The key, then, is to identify physical or chemical properties that most effectively distinguish between lead sources. A large number of physical and chemical measures have been used in fingerprinting analysis (see, for example, D'Haen et al. 2012 or Guzmán et al. 2013). Some use lead-related characteristics to fingerprint lead sources directly (such as lead concentrations, lead-isotope ratios, and ratios of lead concentrations to other elemental concentrations). Other fingerprinting analyses rely on properties of the media with which the lead is associated. For example, the fingerprinting of sediment-associated lead in rivers often relies on the concentration of multiple elements in the sediment or magnetic properties of the sediment in which lead is found.

Once a fingerprint to distinguish the sources has been defined, mathematical models of various types can be used to determine the relative contributions of lead from each source. During the last 2 decades, a wide array of analytic and statistical methods of increasing complexity have been developed to improve source attribution and to characterize the uncertainty inherent in the results.

soil? Are there obvious atmospheric point sources, such as a smelter or mine, in the near vicinity? Has mine-waste material with high lead concentrations been deposited on the site as backfill or roadway material? Is there evidence of storage or disposal of lead-acid batteries on the site? Once basic questions have been considered, a detailed soil sampling protocol can be developed for the specific site. As part of the development of a conceptual model, a handheld x-ray fluorescence (XRF) analyzer can be used to survey the surface-soil lead concentrations and guide soil sampling for additional analyses.

The field characterization should then proceed to determine whether the soil has been disturbed by human activity, whether the site is in an active floodplain and is a location where recent alluvial sediments have been deposited or eroded, and whether the soil was developed in situ (the nature of the underlying parent rock material). Disturbance by human activities is likely to alter spatial and depth trends in lead concentrations and might have led to the direct incorporation of anthropogenic lead into the soil. In Southeast Missouri, for example, coarse-grained tailings (chat) that contain lead were used extensively as fill and thus have been observed to be incorporated into disturbed soils in the region. Similarly, dispersal and source alteration from, for example, weathering processes described in Chapter 3 should be considered.

The distinction between soils composed of alluvial (riverine) and nonalluvial parent materials is an important consideration in site assessment because of differences in vertical patterns of lead concentration. Anthropogenic lead in soils of floodplains that were recently active (in the last century) often extends to depths measured in meters; anthropogenic lead in soils where the parent material was not deposited by the recent action of rivers is generally confined to within a few tens of centimeters of the ground surface. Thus, the topographic setting and type of soil parent material dictate the nature of the sampling program required to assess lead attribution. The following discussion of depth sampling and spatial analysis is for soil developed in place without active deposition where diagnostic vertical distributions tend to be retained as opposed to soils that are subject to recent alluvial deposits (active floodplains).

Spatial and Depth Sampling and Analysis

The spatial distribution of lead in a location is often a clue to its source. Anthropogenic sources of lead in soils are commonly deciphered by using two types of spatial analysis: an assessment of geographic patterns in lead concentration in the topsoil and a spatial analysis of lead as a function of depth below the ground surface. Geographic patterns in lead and other trace-metal concentrations are usually determined by collecting and analyzing composite samples of the topsoil with a gridded sampling scheme (Stafilov et al. 2010) or along transects that radiate from a suspected source, often corresponding to wind direction (Ettler 2016). Alternatively, the data might correspond to the distribution of residences that are sampled in urban areas. With the advent of hand-held XRF units, assessing spatial variation in lead can be easy and rapid and provide key information on lead sources. For example, high concentrations along the dripline of a home suggest that the source is paint or lead particles that have washed off the roof. Materials used for roads or driveways might contribute to high lead concentrations in nearby soils where concentrations decline with distance. Lead distribution, however, might also have nonunique source signals whereby natural, alluvial, or aerially deposited material might all show similar patterns that require further analysis.

Depth sampling and analysis might help to identify the lead concentration in soil parent material and layers that have been enriched by human activities and thus aid in source attribution. Here, the parent material for a soil is simply the rock (or unconsolidated rock material) that now underlies the soil. Soils develop through interaction of parent material with water, gas, living organisms, and organic matter in a given climate over an extended period. Parent material is generally determined through geologic inspection and chemical, mineralogic, and particle-size analysis of samples collected from the bottom of the soil profile and adjacent to the underlying bedrock. To confirm parent composition, multiple elements can be analyzed, and the measured composition compared with the expected weathering behavior of the underlying bedrock derived from modeling (Brantley and Lebedeva 2011). It is gener-

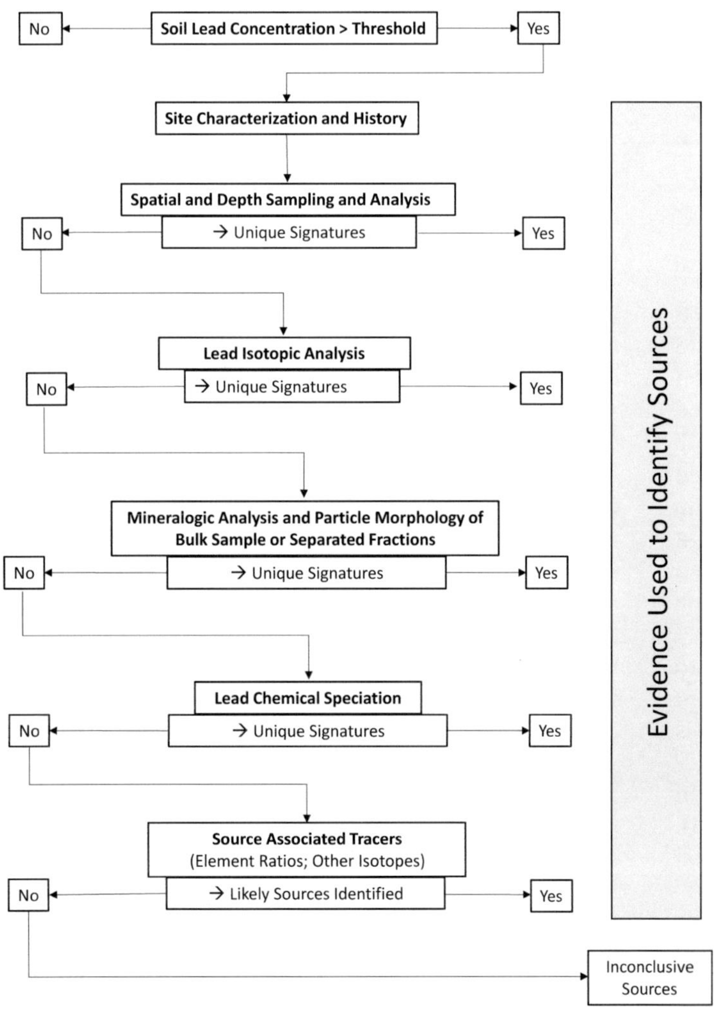

FIGURE 4-1 Example of an investigative strategy for attributing sources of lead in soil. The optimal sequence will depend on site-specific conditions, availability of analytic techniques, and cost and time considerations.

ally helpful to collect multiple samples of parent material if possible to estimate compositional variability, including sampling of local outcrops or drilling material.

Once parent material is identified, the source of lead in the soil can be investigated. Lead that is present in uncontaminated soils derives from the parent material or from natural inputs of dust or sediment. In contaminated soils, lead derives from those sources and from anthropogenic inputs. By sampling and analyzing the depth distribution of lead and, for example, the isotopic signatures, specific minerals, or elemental ratios in the soil, one might be able to determine a scenario that most likely explains the source of the lead. If the physical or chemical characteristics of the lead in the soil are identical with those in the parent material *and* the depth distribution is consistent with this natural source of the lead, the simplest explanation is that all the lead in the soil is derived from the parent material. If the chemical or physical characteristics of the soil lead instead differ from those of the lead in the parent material and its weathering products, the lead has probably been added to the soil. The analytic methods described below help to define the chemical and physical characteristics of the lead to distinguish its source.

In areas of potential ore deposits, separating natural from anthropogenic lead can be complicated. However, for most cases of soils outside a recently active floodplain, anthropogenic lead will generally be concentrated at or near the soil surface and decrease downward to concentrations similar to those of the parent material. In some cases, if topsoil has been placed over mine wastes or chat, vertical distributions might document a layer that has a high lead concentration between layers that have low lead concentrations. Depth sampling of soil profiles can investigate such diagnostic distributions. Such sampling tends to be conducted at uniform increments or to be stratified according to naturally defined soil horizons.

Most commonly, the magnitude of anthropogenic lead is assessed by comparing concentrations in the surface materials with those in deeper horizons, including parent material. The analyses can be complicated by the downward migration of lead, which can occur by leaching (Miller and Friedland 1994; Prapaipong et al. 2008); the translocation of colloids and clay-sized particles (Citeau et al. 2003); biologic activity caused, for example, by earthworms, gophers, ground squirrels, and plant uprooting; and human activities, including gardening and plowing. The depth and rate of downward migration depend not only on the mechanism and timing of mixing but on the chemical form or speciation of lead in the soil; soil characteristics, such as pH, type and amount of organic matter, and clay content and mineralogy; climate conditions, such as rainfall and temperature; and the nature of the particles with which lead is associated (as discussed in Chapter 3).

Interpretation of lead concentrations as a function of depth in soils can also be complicated because even if lead is not added to or lost from a soil, its concentration can increase upward in a soil if high concentrations of other elements have been dissolved and transported out of the soil by natural processes over geologic time. In other words, lead that is not removed from a soil can become more concentrated near the surface when more soluble materials have been removed. To account for weathering, soil scientists generally normalize concentrations of the element of interest, such as lead, to that of another element that is not present in soluble minerals, such as titanium and zirconium. One such normalization, which has been used extensively in the recent literature, is the mass transfer coefficient[1] (Brimhall and Dietrich 1987; Anderson et al. 2002; Brantley and Lebedeva 2011). The mass transfer coefficient of lead is calculated by dividing two ratios. The first is the ratio of the lead concentration in a bulk soil sample to the concentration of an element, such as titanium or zirconium, that is present in the soil as an insoluble mineral. The second is the ratio of those same two elements in the parent material. When the mass transfer coefficient is analyzed as a function of depth, it can reveal patterns of addition or depletion in the soil that can be interpreted in terms of the lead source (Brantley and Lebedeva 2011). Observed depth distributions can be modeled in terms of processes, such as addition, depletion, or biologic cycling of elements into upper soil layers. Using those approaches, several investigators have modeled the addition of lead to soils from atmospheric deposition (Johnson et al. 1995; Drivas et al. 2011), and depth distributions coupled with isotopic signatures of various potential lead sources have been used to determine sources and migration of lead in soils (Prapaipong et al. 2008).

Lead Isotopic Analysis

Measurement of lead isotopic composition can potentially identify a lead source rapidly if the isotopic signature is unique. The effectiveness of isotopic composition as a fingerprinting tool is related to several factors: stable isotopes of lead can be measured with a high degree of precision and accuracy, lead is unreactive and immobile over a wide range of environmental (geochemical) conditions, and fractionation of lead isotopes by physical, chemical, and biologic processes is minimal in contrast with that of stable isotopes of light elements, such as sulfur, carbon, and oxygen (Keinonen 1992; Komárek et al. 2008; Balcaen et al. 2010; Bird 2011; Miller et al. 2012). Thus, isotopic composition of lead of natural and anthropogenic sources (including ore deposits) does not change appreciably as it is dispersed through the environment. In

[1]The use of the term *mass transfer coefficient* by the authors is different from its use in engineering where it is used widely to characterize the transfer rate of mass between phases.

most cases, the variation in a soil will be small relative to the differences between the soil and external inputs; thus, observed differences in the isotopic composition of the soil in question can be attributed to its mixing with lead from other sources (Cheng and Hu 2010). Finally, lead isotopic composition of ore deposits on a global scale is highly variable (Bird 2011; Miller et al. 2015), and lead derived from ore deposits tend to differ from lead typically found in non-ore rocks and soils (Hopper et al. 1991). See Chapter 5 for a detailed case study.

Studies conducted over the last several decades demonstrate that lead isotopes are often effective in determining lead sources in soils if the isotopic values of the source materials are unique (Rabinowitz and Wetherill 1972; Gulson et al. 1981; Steinmann and Stille 1997; Steinnes et al. 2005; Clark et al. 2006; Mihaljevic et al. 2006; Prapaipong et al. 2008; MacKinnon et al. 2011; Ma et al. 2014). In many cases, analyzing the depth distribution and isotopic signatures of lead in the soil allows source determination. If the isotopic character of the lead in the soil is identical with that of the parent rock, the simplest explanation is that all the lead in the soil results from the parent rock material. If, however, the isotopic character of the lead differs from that of the parent material, the lead has probably been added to the soil. As an illustrative example, Ma et al. (2014) completed an investigation in Pennsylvania where lead was enriched in soil relative to the parent bedrock. The isotopic signature of the lead varied significantly and consistently upward from the bedrock to the land surface (Figure 4-2). Using the lead isotopic signature, concentration–depth profiles, knowledge of the history of anthropogenic practices in the area, and models, the researchers attributed the added lead to atmospheric deposition between the 1850s and 1920s, when lead was emitted to the atmosphere from coal-burning and ore-smelting in the area (Ma et al. 2014).

Similarly, Steinnes et al. (2005) examined lead concentrations and sources throughout Norway. Their findings illustrate a similar $^{206}Pb/^{207}Pb$ ratio for C-horizons (unconsolidated parent rock material) throughout the country and are indicative of the undisturbed soil baseline lead signature. Using known $^{206}Pb/^{207}Pb$ ratios in atmospheric deposition, they were able to determine the fraction of the lead attributed to atmospheric inputs for lead-enriched surface soils. Gulson et al. (1981) provide another illustrative example. They examined lead sources

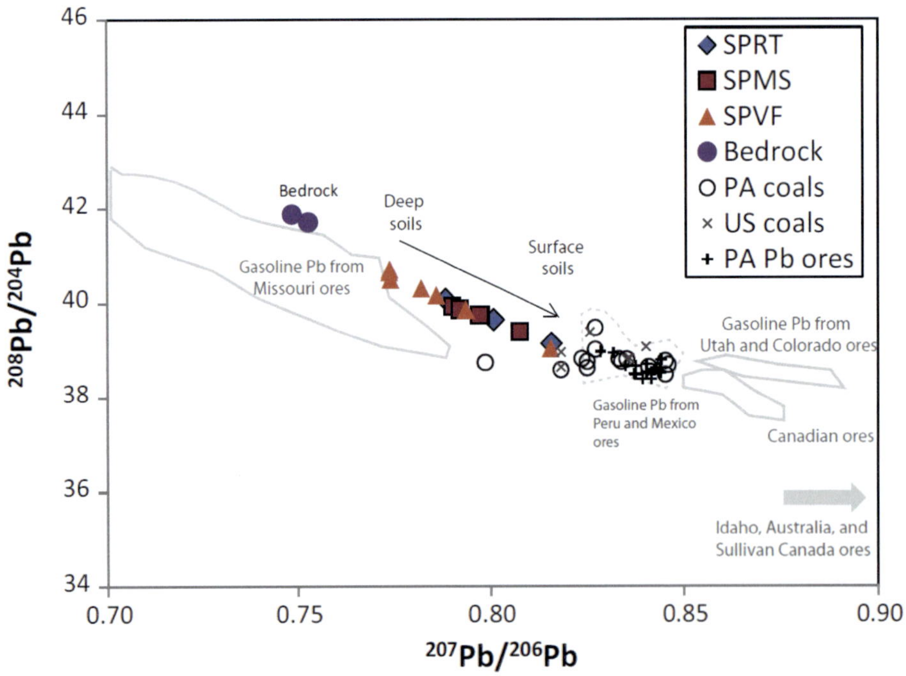

FIGURE 4-2 Isotopic signatures of soil samples from the pristine Shale Hills site in central Pennsylvania compared with various sources of lead. The gray circled areas represent lead-isotope ratios for leaded gasoline *if* the lead were derived from the ores indicated. Solid-colored symbols represent soil or bedrock samples, open symbols represent samples of coal from Pennsylvania, and crosses represent lead ore mined in nearby parts of Pennsylvania. Dashed gray lines indicate the isotopic values of Peruvian and Mexican lead ores that were generally not used in the Great Lakes region (Graney et al. 1995). Soil lead is attributed to a mixture of bedrock and atmospherically deposited lead from local coal-burning and iron or lead ore-smelting; deeper soils contain more bedrock-derived lead, and topsoils contain more coal- or ore-derived lead. Abbreviations: SPRT, southern planar ridge top; SPMS, southern planar middle slope; SPVF, southern planar valley floor. Source: Ma et al. (2014). Reprinted with permission; copyright 2014, *Anthropocene*.

in soils near Adelaide, South Australia, by comparing concentrations and isotopic signatures between surface (upper 10 cm) and subsoil (30-40 cm) collected along a path that included urban and rural land use. Subsoil isotopic values were similar throughout the traversed sampling path and were consistent with parent rock; lead in the subsoil was therefore attributed almost entirely to that derived from the parent rock. In contrast, lead from parent rock made up a minor fraction of the total lead in surface soils. Using the isotopic signatures along the transect with the depth comparison, the authors were able to eliminate pesticides, power stations, and smelters as major sources of soil lead; instead, they attributed the lead to gasoline.

Rabinowitz and Wetherill (1972) used isotopic signatures, lead concentrations, and general mineralogic observations to assess sources of soil lead in shallow soil samples (0–2.5 cm and 12.5 cm) in Southeast Missouri and Benicia, California; their work illustrates the benefits and limitations of isotopic analysis for source attribution. In the majority of sites examined from St. Louis to rural areas away from any known ore bodies and sites near ore-processing or gasoline-refining, lead enrichment in surface soil layers could be attributed to specific sources on the basis of comparison with deeper soil layers and measured isotopic values of potential sources. However, a slight enrichment of 25-ppm lead in a surface soil layer near the ore-crushing and beneficiation plant in Indian Creek Mill, Missouri, provided an equivocal isotopic signature. In that case, mineralogy, particle-size distribution, and lead isotopic signature were not sufficient to resolve the source of high lead concentrations in the surface soil. In the other 11 sites examined in Southeast Missouri and Benicia, California, the combination of lead isotopic signature, concentration profiles, and mineralogic analysis allowed source attribution of lead enrichment in surface soil layers.

Using spatially resolved depth variation in lead concentration and lead isotopic signatures, Prapaipong et al. (2008) determined lead sources for soils potentially affected by smelting in Southeast Missouri. Lead isotopes of the smelting ore, which originated from the Viburnum Trend, were distinct from those of the parent materials of soils developed on Bonneterre dolomite or St. Francois rhyolite, and they also differed from the isotopic signature of lead derived from burning of leaded gasoline. The unique signature of each potential lead source allowed apportionment across the study area and with sampling depth. Lead concentrations attributed to smelting decreased with depth to a consistent value (typically 10–30 ppm) and decreased with distance from the smelter. However, within 2 km of the smelter, up to 90% of the lead was derived from smelting operations to depths of 25 cm. Over 4 km away from the smelter, smelter emissions accounted for 10–50% of the soil lead in the upper 15–25 cm. Isotopic analysis revealed that aerially deposited lead (from smelting and automotive exhaust) was present even at depths where concentrations were consistent with natural background. That finding suggests that the naturally occurring lead was displaced from the soil and that a maximum retention capacity was reached; any new lead that entered the soil would therefore either move through the soil or displace lead already present in the soil. To establish temporal trends in soil lead inputs, Prapaipong et al. (2008) examined lead concentrations and isotopic signatures of White Oak tree cores by using tree rings to provide concentration–time profiles. Lead concentration rose sharply 7 years after onset of smelting operations, increased within the tree rings over the following 25 years, and increased by a factor of 10 in trees within 1 km of the smelter. Isotopic analysis substantiated that the rise in lead concentrations was from the uptake of smelter-derived lead. The combined use of tree cores to determine temporal trends with spatial and depth profiles of soils to determine lead concentrations and isotopic signatures provides a powerful means of identifying lead sources.

Thus, if the lead isotopic compositions of the potential sources are unique and the lead isotopic values in the soil of concern fall within the range of the source materials, the contribution of each source to lead contamination can be determined. Given the robust nature of the technique, it might obviate additional analyses for source attribution, and this would greatly reduce the time, effort, and costs involved in determining lead sources. If the lead isotopic compositions of the source materials are similar (that is, they overlap), the method cannot unequivocally assess lead sources, and additional techniques are required.

Mineralogic Analysis and Particle Morphology

Mineralogic analysis of soils, such as optical microscopic identification of sand particles and x-ray diffraction (XRD) analysis of silts and clays, might provide signatures that allow source attribution. Minerals unique to specific sources can help with source attribution. The presence of galena, PbS, and other heavy minerals, for example, might be a clear signature of lead source (Ettler et al. 2005; Cabala and Teper 2007), whereas hydrocerussite, $Pb_3(CO_3)_2(OH)_2(s)$, might precipitate as a soil mineral after weathering of the original lead mineral. As lead disperses in the environment, soil concentrations tend to decrease as distance from the source increases, and lead is more likely to be adsorbed to other soil minerals or to be present as a minor substituent in secondary phases in soils and sediments. As a result, source attribution through mineralogic analyses, although often effective, can be inconclusive in other cases.

XRD has been used extensively for clay mineral identification (see Whittig and Allardice 1986) and to elucidate soil and sediment sources (Eberl 2004; Benedetti et al. 2006). It is often used in conjunction with particle-size or density separation to identify minerals specific to a lead source (Olson and Skogerboe 1975). The mineralogy and particle structure determined with optical microscopy and scanning electron microscopy (SEM) of sand particles has been used extensively on alluvial and dust deposits to track sediment source (Arribas et al. 2000; Abu-Zeid et al. 2001; Cardona et al. 2005; Benedetti et al. 2006) and might serve to attribute lead sources in soil.

SEM coupled with elemental analysis—conducted with energy dispersive x-ray spectroscopy (EDS) or wave-length dispersive x-ray spectroscopy—is an additional means of obtaining chemical information and information on particle structure and texture. A description of the attributes and limitations of SEM and elemental analysis for identifying particle sources is provided in Pye (2004). Specific elemental ratios and morphologic characteristics of particles can be used to infer mineral identity, and combining mineralogy, particle size, and elemental composition has proved successful in source attribution (de Boer and Crosby 1995; Sterling et al. 1998). For example, Linton et al (1980) used specific particle structure and chemical composition to distinguish lead sources between automotive exhaust and paint chips, and Sobanska et al. (1999) used spherical particle structure and elemental composition to identify lead originating from smelters. Demonstrative features, such as spherical structure associated with smelters, can provide definitive source identification, and more subtle etching of sand grains might provide important clues to particle source (Abu-Zeid et al. 2001; Cardona et al. 2005). Figure 4-3 illustrates the use of SEM and EDS to distinguish between spherical lead-containing particles that are indicative of high-temperature processing, such as smelting, and particles that are more characteristic of airborne dust originating from vehicle activity or wind. Transmission electron microscopy (TEM) coupled with EDS and electron diffraction might also be useful for source identification. Mineralogy and particle-surface morphology can be deduced for isolated particles with TEM, which might be particularly useful for small, well-crystallized particles that have a propensity for deposition via wind-borne transport (Buseck and Posfai 1999).

Lead concentrations also tend to depend on particle size in material deposited as dust, with larger particles deposited closer to the source (NEPC 1998). Thus, particle-size analyses that include dividing particles into fine and coarse fractions can help to determine source. Size separation is typically conducted with wire-mesh sieves for particles larger than about 0.05 mm; finer particles are measured by sediment rate separation, suspension density (hydrometer), or laser light scattering. An example of particle-size variation is illustrated by Taylor et al. (2010), who found that lead concentrations within 2 km of mining and smelting activities in Mount Isa, Queensland, Australia, were higher above more distant soils and depended on particle size. High lead concentrations associated within finer particles (<180 μm) extended further from the site. Such geographic concentration data are often combined with dispersion models that predict the downwind deposition and surface concentrations of atmospheric lead from known sources (see, for example, Small et al. 1995).

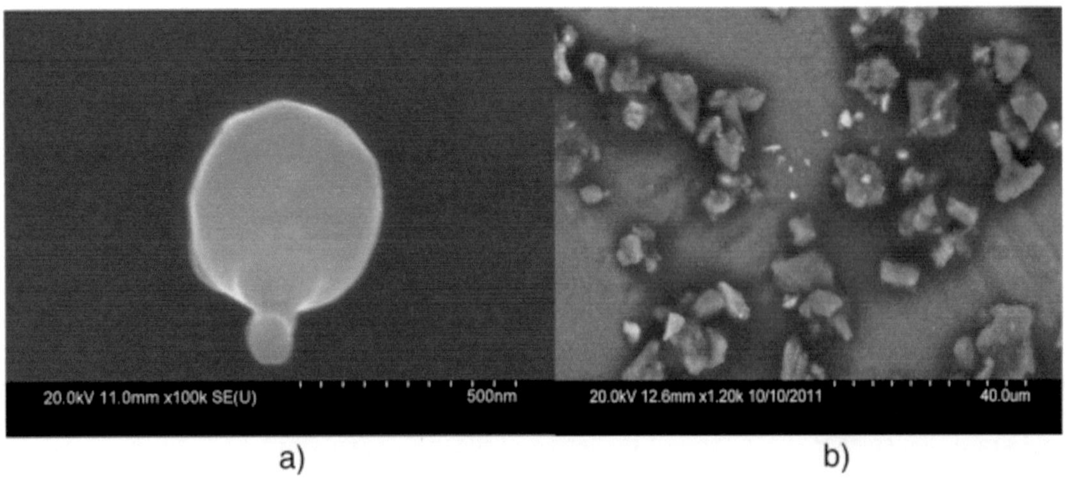

FIGURE 4-3 Representative SEM images of particles with aerodynamic diameters of (a) 0.32 μm and (b) 6.2 μm collected from a micro-orifice uniform deposit impactor at an active copper-smelting site (Hayden, AZ). Particle depicted in (a) was found with EDS to contain 10.88 ± 1.00% lead by weight. Source: Csavina et al. (2014). Reprinted with permission; copyright 2014, *Science of the Total Environment.*

Heavy minerals—such as pyrite, galena, cerussite, anglesite, and sphalerite—can be separated from soils by using fluids of different density. Such fractionated samples (separates) can often be used to associate soil grains with sources of lead contamination. Heavy minerals are particularly effective in tracing airborne lead and other metals derived from smelting and mining operations (see review by Ettler et al. 2016). Heavy mineral assemblages are often examined in combination with particle structure for attribution purposes. For example, the combination of ilmenite, $FeTiO_3$, and particle structure allowed successful determination of particle origin (Basu and Molinaroli 1991), and heavy minerals with quartz stress fractures provided evidence on wind-transported particles (Abu-Zeid et al. 2001).

If heavy minerals are isolated with density separation methods that rely on high-viscosity fluids, the assemblages can be characterized by using mineralogic tools noted above, such as XRD (Burt et al. 2003; Ettler et al. 2005) and SEM–EDS (Cabala and Teper 2007). Density separation has been used extensively for the analysis of minerals in both consolidated and unconsolidated rocks (see, for example, Morton 1985), but it has not been applied much to source attribution of anthropogenic lead in soils. Nevertheless, the method holds promise in that it allows additional mineralogic, geochemical, or isotopic analysis (described above) to be performed directly on heavy minerals.

The attribution of airborne lead and other metals to mining and smelting operations by using heavy minerals has generally been based on spatial variations in the quantity with which they are found in soils relative to the source and the area's predominant wind directions. Calaba and Teper (2007), for example, found that the proportion of heavy minerals, particularly iron sulfides associated with ore minerals, reached concentrations as high as 6–8% within several kilometers of ore-processing sites in southern Poland. Similarly, Ettler et al. (2005) found that anglesite occurred commonly in the dust and fly ash of lead-smelting operations in the Příbram district of the Czech Republic. Thus, they were able to argue that their presence in local soils was a clear indicator of airborne contamination from the nearby lead smelters.

Lead Chemical Speciation

Lead speciation can provide additional means for source attribution, but chemical speciation of lead in soils can be challenging because lead can be dispersed or poorly crystallized or be associated with small particles, all of which makes it difficult to conduct XRD or SEM analyses. Furthermore, weathering, alteration, and dissolution of primary lead compounds can change lead speciation and often result in adsorption of lead on mineral or organic-matter surfaces or its substitution as a minor element in newly precipitated phases. To overcome those challenges, selective extractions have been used commonly to assess the chemical speciation of lead.[2] Most series of selective extractions are based on a method developed by Tessier et al. (1979); various methods are reviewed and summarized by Ure et al. (1993). One potential problem is that although the extracting medium might target lead associated with specific materials (such as organic matter, carbonates, or iron or manganese oxide), it might also capture other material; that is, the separation is not clean. Coupling that problem with changes in the lead species leads to imprecision in the method. The accuracy of a series of selective extractions has been assessed by comparing them with direct spectroscopic analyses—such as x-ray absorption spectroscopy, x-ray fluorescence, Raman spectroscopy, and other advanced analytic methods (Scheckel et al. 2005; Hayes et al. 2009; Hayes et al. 2012)—that provide information on chemical bonding of lead. In a study of mine tailings from Leadville, CO, coupling of direct characterization of lead speciation that used x-ray absorption spectroscopy with a series of selective extractions revealed discrepancies between lead fractions targeted in extractions and lead speciation determined spectroscopically and demonstrated substantial redistribution of lead after extraction treatment (Ostergren et al. 1999). However, despite their limitations, selective extractions are useful for assessing the relative reactivity of lead in different soil fractions that can then be compared with other soil layers or other sites and sources. For example, particle-size separations and selective extractions were used with x-ray absorption spectroscopy and x-ray fluorescence to determine lead speciation and mineral phases in core samples from weathered mine tailings to assess the potential for wind transport of particles and their relative bioaccessibility (Hayes et al. 2012). More advanced spectroscopic methods for chemical speciation might also be useful but can be difficult to access or can require specific expertise for data acquisition or analysis; a comprehensive summary of potential techniques is provided in D'Amore et al. (2005).

Source-Associated Tracers

The final source-attribution approach for soils encompasses a wide array of methods, including multielement analysis and methods that involve stable isotopes other than lead; they have been used to determine the contribution of lead from multiple sources on the basis of composition of the geologic materials with which lead is associated. The methods hold considerable promise but have not been extensively applied to the specific problem of lead-source attribution. For example, cadmium and zinc are commonly associated with and recovered during the

[2]Selective extraction is an analytic process in which lead is leached from a sample with a solution to recover only a specific form of lead.

processing of lead ores (Rehkämper et al. 2011). As the ore is smelted, light isotopes of zinc and cadmium preferentially exit with the exhaust; heavy isotopes of zinc and cadmium are preferentially retained in the smelting residue (Sivry et al. 2008; Mattielli et al. 2009; Miller et al. 2015). The differences might then be used to track the deposition of emitted smelter particles, including lead-containing particles, over the region by analyzing the zinc and cadmium isotopic composition of the soils (see, for example, Cloquet et al. 2006; Mattielli et al. 2006, 2009; Shiel et al. 2010; Thapalia et al. 2010). Similarly, stable zinc and copper isotopes were used by Thapalia et al. (2010) to identify smelter emissions in lake sediments and thus trace lead to the smelter emissions.

In other cases, elemental ratios have been an easy means of tracing lead sources. Zota et al. (2011) used the association of lead, zinc, cadmium, and arsenic to identify particles originating from chat, and manganese was used to identify particles from uncontaminated soil. To elucidate whether pesticides or paints were responsible for enriched soil lead, Nezat et al. (2017) used the As/Pb ratio in particles to denote pesticide contributions and Pb/Ba/Zn ratios for paint contributions. Given the use of antimony in alloying processes at specific smelters, Eckel et al. (2002) found Sb/Pb ratios, as opposed to As/Pb or Zn/Cd ratios, to correlate with soil lead contributions from smelting operations.

Analytic methods have also been paired with modeling to attribute sources, and the following discussion provides several examples of such approaches for source attribution. An example of combined analysis and modeling is the use of SEM–EDS to develop a hierarchical cluster analysis for household-dust apportionment based on particle size and element ratio (Hunt et al. 1992). A combination of multiple analytic methods, extractions, and geospatial modeling was used to characterize iron oxide phases, fine-grained mine waste, and organic material that adsorbed lead (and other metals) and to distinguish relative contributions from two geochemically distinct sources, mine drainage and mine-waste pile runoff (Schaider et al. 2014). Simple mass-balance models—an example of which is provided in Chapter 5—can also help to constrain potential source contribution and aid in source attribution. More sophisticated predictive statistical models can also be useful in lead-source attribution. Contributions of soil lead from an automotive-battery recycling facility were compared with lead-based house paint and other background sources by Small et al. (1995); they used scaled deposition estimates from atmospheric dispersion to develop a statistical model that characterized observed soil-lead concentrations and attributed sources relative to the urban background. Multielement models noted in the next section on sediments might also be applicable to lead-source attribution in soils.

SEDIMENTS

The committee's investigative strategy for sediment is shown in Figure 4-4 and provides a guide to the effective determination of the natural and anthropogenic sources of lead found in river sediments. Similar analytic methods can be applied to nearly any type of sediment, such as those associated with lakes, wetlands, peat bogs, and coastal or estuarine environments. In fact, it is common to analyze sediments from different types of settings to provide a more holistic assessment of inputs of lead to the environment. Nonetheless, the focus here is on rivers and river sediments.

The sequence of analytic methods is intended only as a guide and might need to be adjusted to fit the site or operable unit of interest given specific site conditions, availability of instrumentation, and considerations of cost and time (see Box 4-2). The discussion that follows provides a general description of the methods shown in the figure and highlights some advantages, disadvantages, and sources of uncertainty. Additional information on each method can be obtained from the references provided.

Site Characterization and History

As applied here, site characterization is the process of collecting data to describe the river basin in terms of its physical, hydrologic, physiographic, and biotic characteristics at the time of the analysis and in the past. The exact list of measures to be used will depend on the site conditions but should include the identification of all natural and anthropogenic lead sources in the site and in the basin upstream of the site. The objective is to document the location and spatial distribution of each source, the form of the lead and associated materials in each source, the pathways and processes through which lead could enter a river from each source, and, for anthropogenic sources, the history and methods of lead use or processing. Most of the data required for site characterization will probably have been collected through the Superfund remedial investigation and feasibility study.

Fluvial Geomorphic Analysis and Sampling

Rivers can be viewed, simplistically, as networks of channels that transport water and sediment from upstream areas to downstream areas of a basin. Rather than being simple conduits, however, they are highly dynamic features in which water is continuously eroding and depositing sediment in the river channel and on the floodplain during floods. Erosion and deposition of sediment can incorporate sediment-associated lead that enters the drainage network into riverbed and floodplain (bank) deposits. The deposits vary widely along the river valley in their areal extent, age, and sedimentologic characteristics, such

BOX 4-2 Examples of Modifications of Investigative Strategy

Figure 4-4 is provided as a guide to the effective attribution of lead in river sediments in or adjacent to Superfund sites. The investigative strategy will ultimately depend on a host of factors, including the nature of the lead sources, the size of the area under investigation, the available analytic capabilities, and constraints on time and money. Thus, the strategy will probably need to be modified by selecting methods that possess a high probability of being effective for a specific site or operable unit. For example, the Maritsa River in Bulgaria is highly contaminated by lead and other trace metals. Contamination of the river is related to the inflow of mining and smelting waste from multiple mining districts in its tributaries. Bird et al. (2010a) demonstrated that the various ore bodies in the basin and the mine wastes associated with the ore bodies exhibit a distinct lead isotopic signature. As a result, lead-isotope data could be used to apportion lead between the mine or smelting wastes and the local bedrock. Lead isotopes could not, however, be used to apportion lead in riverbed sediments of the larger Maritsa River because lead was derived from many more sources, including multiple ore bodies and bedrock types. The occurrence of multiple lead sources precluded the use of simple mixing models. Bird et al. (2010a) circumvented the problem by determining the relative percentage of sediment in the Maritsa River derived from each tributary and using multi-element fingerprinting. By treating, the tributaries, including those contaminated by mining operations, as distinct sources of sediment and sediment-associated metals, the authors were able to estimate the relative quantity of sediment and, by association, metals contributed to the Maritsa River by each tributary. They were also able to estimate changes in tributary contributions downstream along the river.

The Rio Loa, which is in the Atacama Desert of northern Chile, provides another methodologic example (Miller et al. 2013a). The Rio Loa is highly contaminated by lead and other metals and metalloids. The predominant sources of metal contamination include large copper mines and their tailings piles in the Rio Loa basin and the El Tatio geyser basin. Lead isotopes could not be used to differentiate between lead from the mines and the geyser basin, although they could be used to distinguish between lead from the mines or geysers and non-mineralized rocks in the area. Current studies suggest that the most effective method for sediment and sediment-associated metal apportionment is the ratio of copper (which is high in mining wastes) to antimony (which is high in sediments of the geyser basin). The abundance of heavy minerals also appears useful.

as particle size, shape, lithology, and color. The sources of sediment and lead in the deposits also vary along the river and on a local (stream segment) scale.

The primary objective of analyzing a river's erosional and depositional history is to identify, characterize, and map the distribution of distinct, time-correlated sedimentary deposits along the river valley. The analysis provides an understanding of the transport pathways through which sediment-associated lead is dispersed from suspected lead sources and the potential sites where lead might have accumulated along the river during specific times. It also provides a framework that is essential for developing a sound sampling program for which the physical and chemical analyses described below can be effectively conducted. For example, sediments deposited before European settlement might provide information on natural concentrations of lead associated with the sediments in the river basin.

Lead Isotopic Analysis

When applied to river sediments, the assumption is that lead in riverbed, floodplain, or terrace deposits represents a mixture of lead from all the sources that contribute lead to the river at the time of sediment deposition. Comparison of the lead isotopic composition of the sediment mixture with the isotopic composition of the suspected sources allows an estimation of the relative contribution of lead to the sediments from each potential source. The analysis is often complicated because the lead in river sediment is derived partly from the bedrock that underlies the basin. Depending on the basin's geology, the lead isotopic composition of the underlying bedrock might vary substantially between rock units. Thus, in addition to the suspected anthropogenic sources of lead, it might be necessary to consider multiple bedrock lead sources.

Nonetheless, many studies have used lead isotopic methods effectively to identify the source of lead in river deposits directly (Church et al. 2004; Miller et al. 2007; Shepherd et al. 2009; Bird et al. 2010a,b; Ayrault et al. 2012). Lead isotopic methods have also been applied to various other sediment types, such as those associated with reservoirs and lakes (Shirahata et al. 1980; Petit et al. 1984; Graney et al. 1995; Chiaradia et al. 1997; Foster et al. 2002; Callender 2005), wetlands, and peat bogs (Shotyk et al. 1998; Weiss et al. 1999; Marcantonio et al.

2000; Cloy et al. 2008). Lead isotopes have been particularly useful in deciphering the contributions of lead from nonmineralized rocks and mining and milling wastes and contributions of lead from distinct mining districts when the lead isotopic characteristics vary between ores (see, for example, Bird et al. 2010a,b). However, when the isotopic characteristics of the mining wastes are similar to those found in the local, mineralized, but unmined bedrock, it will not be possible to attribute lead to those sources by using lead isotopes.

The analysis of river deposits differs from the analysis of upland soils (described in the previous section) in two ways. First, the isotopic composition of river deposits of known age can be determined and compared with suspected lead sources in the basin for that time. In doing so, it might be possible to construct a history of the changes in lead source contributions to the river over periods of decades to centuries. Second, spatial variations in lead concentration and isotopic data can be combined to assess changes in lead source longitudinally along the river valley. For example, Shepherd et al. (2009) analyzed a variety of geologic materials along an 86.5-km segment of the River Wear in northeast England (Figure 4-5). At Site 1, the $^{208}Pb/^{206}Pb$ ratio in riverbed sediment was comparable with that in geologic materials found in a mineralized ore body and was therefore interpreted as reflecting the relatively high lead concentrations associated with historical mining operations. The decline in downstream concentrations between Sites 1 and 5 was thought to result from dilution and the deposition of dense, finely milled lead particles in the riverbed. Farther downstream, $^{208}Pb/^{206}Pb$ ratios continue to decline until reaching Site 9, whereas lead concentrations increased. The decrease in $^{208}Pb/^{206}Pb$ ratios presumably reflected the influx of mine waste from two tributary valleys where lead ore was derived from a second mineralized zone in the ore body; this second mineralized zone was characterized by a lower $^{208}Pb/^{206}Pb$ ratio. Shepherd et al. (2009) used changes in lead concentration and $^{208}Pb/^{206}Pb$ ratios between Sites 10 and 27 to assess the influx and importance of additional sources of

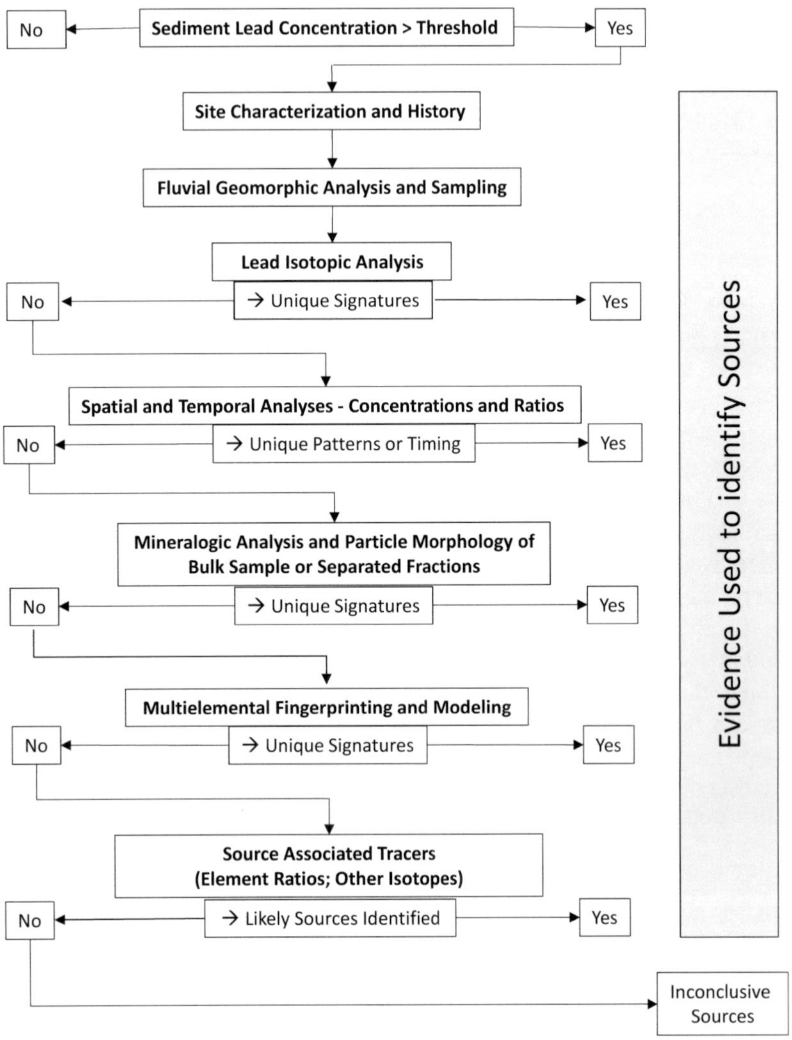

FIGURE 4-4 Example of an investigative strategy for attributing sources of lead in river (alluvial) sediments. The optimal sequence will depend on site-specific conditions, availability of analytic techniques, and cost and time considerations.

lead to the river. The key point is that combined spatial analysis of lead concentration and lead isotopic composition of the river sediments might provide a more detailed understanding of source contributions of lead to a river than is possible by using lead concentration data alone.

Spatial and Temporal Analyses

If the use of lead isotopes fails to determine the source of lead in river sediments completely, it will be necessary to apply one or more additional methods. Spatial and temporal analyses have been effective in determining lead sources qualitatively in many river basins and are described below.

Spatial Analysis

One of the most extensively used methods for determining sources of lead in rivers is examination of the spatial distribution of lead concentrations or elemental ratios, such as Pb/Zn, in alluvial deposits. The initial interpretation is that an abrupt downstream increase in lead concentration or the ratio of lead to another element will correspond geographically to a predominant source. There are several advantages to the approach, including that it can be applied to both riverbed and floodplain deposits, it can be completed relatively rapidly by using samples collected during a single sampling campaign, it relies on standard analytic procedures for the analysis of total lead concentrations, and the data are relatively easy to interpret.

An example was provided in Chapter 3 (Figure 3-3) for lead, zinc, and antimony concentrations along the Rio Chilco–Rio Tupiza drainage system of southern Bolivia (Villarroel et al. 2006). In the United States, Integral (2014) and Pavlowsky et al. (2010) used spatial variations in lead concentrations in channel sediments along the Big River to assess lead sources. Figure 4-6A shows that lead concentrations increase abruptly downstream of river mile 110, presumably in response to the influx of mining-related waste materials as the river passes through the St. Francois County Mining Area. Lead concentrations decrease downstream until reaching the Meramec River. However, the general decrease in concentration is disrupted locally by minor increases in lead concentrations (Figure 4-6A). Integral (2014) argued that those minor increases in lead

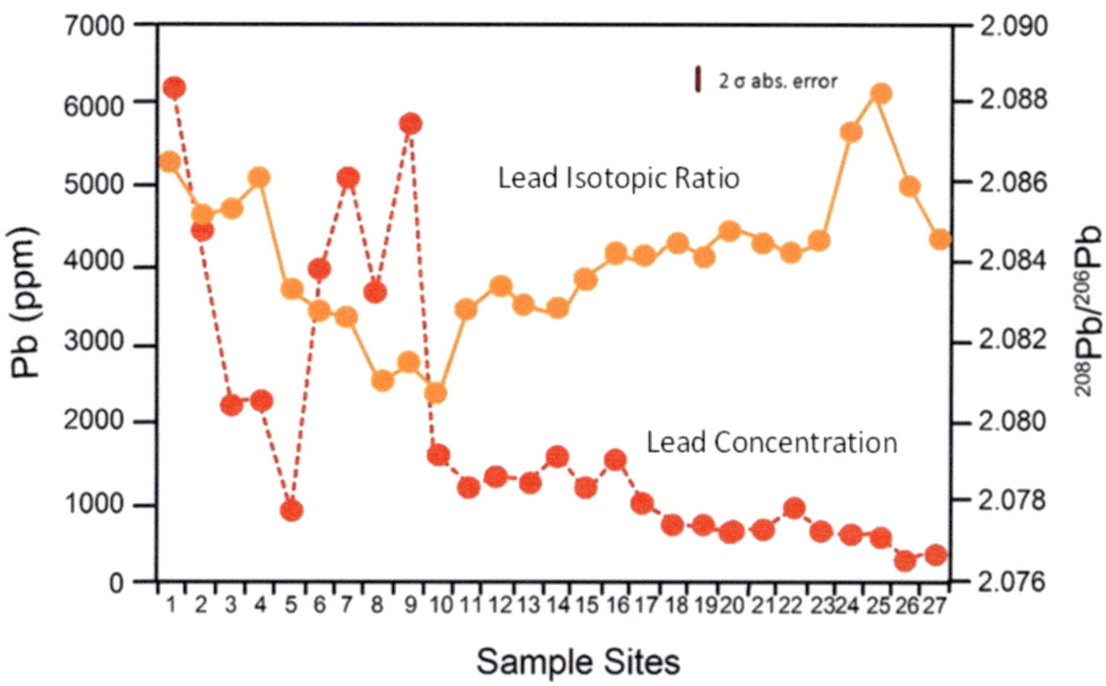

FIGURE 4-5 Downstream variations in total lead concentrations and $^{208}Pb/^{206}Pb$ ratios along the Wear River in northern England. Total distance is about 86.5 km. Error bar refers to isotopic measurement. Source: Shepherd et al. (2009). Reprinted with permission; copyright 2009, *Science of the Total Environment*.

concentration downstream of river mile 85 reflect less important additions of lead from other mining operations, including lead from the Barite Mining District.

Although the analysis of spatial patterns in lead concentrations and elemental ratios provides insights into the sources of lead in river sediments in the drainage basin, it has several limitations as described below.

• Multiple sources might exist in a specific area, and it might not be possible to differentiate inputs between multiple overlapping sources by using lead concentration data alone. For example, the input of mining wastes might spatially correspond to areas of mineralized rock where relatively large quantities of natural lead might also enter the channel. Exploration geochemists have applied similar spatial analyses to rivers to identify the location and grade of ore deposits. To help alleviate the problem, downstream variations in other elements or in lead ratios (such as Pb/Zn) might be used to gain additional insights into the source of the lead if the concentrations of such elements differ between the overlapping sources. Figure 4-6A, for example, shows that barium concentrations increase substantially downstream of river mile 85; a trend that Integral (2014) argues is related to the influx of barite from mining operations in the Barite Mining District. Alternatively, the analysis of downstream trends in elemental concentration might focus on a specific sediment size fraction, as shown in Figure 4-6B, which is associated with a known lead source, such as tailings.

• The resolution of the downstream trends in lead concentrations reflects the number of samples that are collected along the river and the location from which the samples were obtained. Differences in sampling frequencies and locations can yield differing results. For example, Figure 4-6 shows the downstream trends in lead concentrations developed by Intergral (2014) (Figure 4-6A) and Pavlowsky et al. (2010) (Figure 4-6B). Differences in the trends in lead concentration are related primarily to differences in the sampling scheme and approach used, which lead to differences in interpretation of the primary source of lead to the Big River.

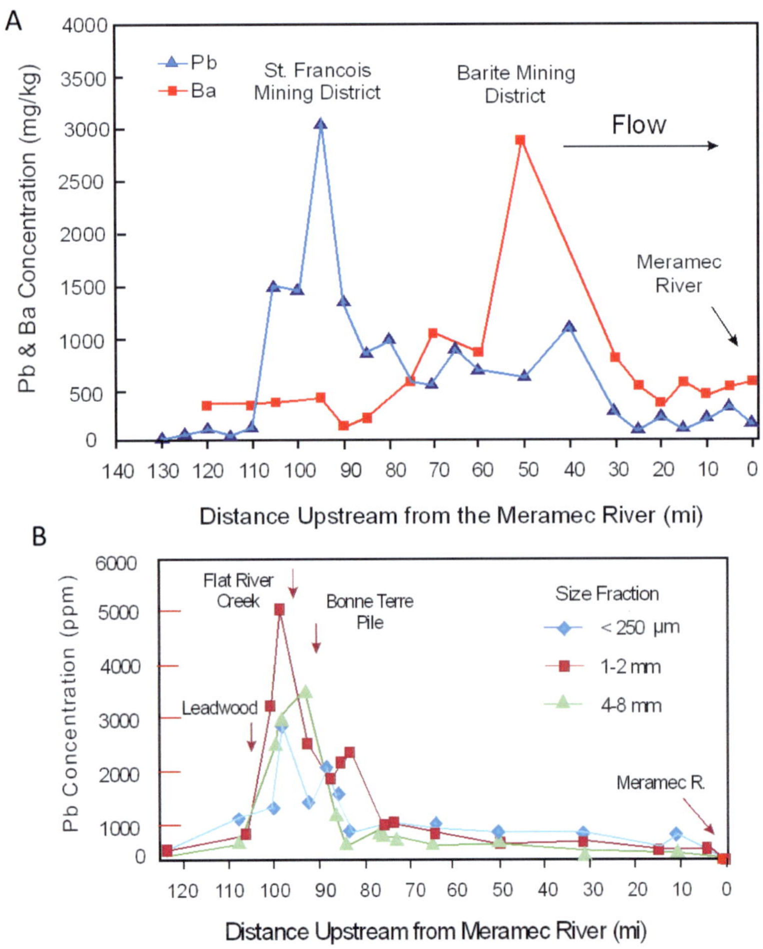

FIGURE 4-6 (A) Changes in lead and barium concentration along the Big River as determined by Integral (2014). Abrupt downstream increase in lead concentration has been interpreted as reflecting the input of mine-associated lead in the St. Francois Mining District. Increased barium and minor increases in lead presumably reflect the input of sediment from mining wastes in the Barite Mining District. Source: Adapted from Integral (2014). (B) Changes in lead concentration along the Big River as determined by Pavlowsky et al. (2010); data are stratified by grain size. Source: Adapted from Pavlowsky et al. (2010). Differences in developed trends partly reflect differences in sampling frequency and location.

• Abrupt downstream variations in lead concentrations might reflect erosional and depositional patterns along the river rather than a location of lead influx. Stream segments that are characterized, for example, by relatively wide valleys, low gradients, and gentle flows are often areas of sediment-associated lead deposition. That deposition can lead to zones of high lead concentrations; thus, abrupt increases in lead concentration are not always a perfect indicator of a lead source.

• The method cannot be used to determine the quantity of lead inputs into a system from a specific source. Rather, it provides a means of assessing the relative importance of lead inputs from a given source.

Temporal Analysis

Floodplains and terraces are composed of sediments that are deposited directly by the river adjacent to the channel (Figure 3-4). In contrast with upland areas, anthropogenic lead in floodplains and terraces often extends to depths measured in meters. The spatial distribution of sediment-associated lead in these alluvial deposits reflects the depositional processes responsible for floodplain formation and the sources and concentrations of lead in the river at the time of deposition. The floodplain and terrace deposits therefore contain a historical record of lead inputs (Macklin et al. 1999, 2006). Once deciphered, this stratigraphic record can be compared with documented variations in lead-related human activities to help to determine the potential role of anthropogenic lead sources in contaminating the river environment. The lead-related human activities should be described in terms of their time of operation, spatial extent and location, quantity of potential lead inputs into the river, and the geochemical and physical character of the contaminated sediments (Heim and Schwarzbauer 2013), all of which are to be collected as part of the site characterization described above.

The methods, assumptions, and uncertainties inherent in developing a chronology of trace-metal inputs (including lead) into a river are outlined in Macklin and Klimek (1992), Hudson-Edwards et al. (1997), Tobin et al. (2000), Miller and Orbock Miller (2007), Pease et al. (2007), Du Laing et al. (2009), Bábek et al. (2011), Matys Grygar et al. (2011, 2012, 2013), and Ferrand et al. (2012). In general, it involves documenting the physical characteristics of the river deposits, sampling and analyzing the deposits for lead and other substances, and applying absolute dating methods to estimate the age of the deposits in calendar years. Absolute dating methods include a wide array of techniques, such as measurement of radioactive isotopes (such as ^{137}Cs, ^{210}Pb, and ^{14}C), optically stimulated luminescence, dendrochronology, and identification of historical artifacts; see Stokes and Walling (2003) for a more detailed discussion of these and other dating methods.

The most detailed assessment of lead and trace-metal attribution is based on the analysis of overbank deposits. Overbank floodplain deposits—also referred to as vertical accretion deposits—are produced by the deposition of suspended sediments from floodwaters that inundate the floodplain surface (Figure 3-4). The process results in layers of younger materials that bury older sediments; thus, the sediment age varies systematically with depth. Because deposition occurs only during floods, there are temporal gaps in the depositional record, and the relationship between depth and unit age is not perfect. Nonetheless, such depositional sequences often result in the construction of a relatively high-resolution history of lead loadings into the river.

The distance of a specific stream segment from a source and the number of potential upstream lead sources vary along the river valley (Figure 4-7). Therefore, it is important to develop a chronology of lead inputs into the river at multiple locations. Sites should be selected to ensure that all potential lead sources will be reflected in the constructed chronology and to minimize the number of locations needed to differentiate inputs between all potential lead sources. The historical record should also extend back hundreds to a few thousand years, if possible, to ensure that natural variations in lead loadings (those characteristic of the basin before substantial human activities) are adequately characterized.

A potential disadvantage of the approach is that multiple human activities that have the potential to increase lead influx in a river can overlap temporally. For example, early mining operations in Southeast Missouri during the late 1800s and early 1900s correlate temporally with regional land-use change, which allowed large quantities of eroded upland soil to enter the drainage network. Increases in sediment-associated lead during that time might therefore have been associated with both mining operations and regional erosion of mineralized soil that contained natural lead. In some cases, it might be possible to separate the significance of lead from such diffuse sources from that of point sources by examining spatial trends in the data.

The method also assumes that lead in the sedimentary deposits is immobile, an assumption that is typically met (Hudson-Edwards et al. 2001). Provided that the migration of lead after deposition is negligible, the resolution of the record depends on the degree to which the sediments can be dated; when the ages of the sediments are poorly constrained, only general statements about lead sources and contributions can be made. Other causes of uncertainty are associated with spatial differences in the deposition of overbank sediments across the floodplain, variations in concentration related to flow conditions at the time of sediment deposition, and variations in the grain-size distribution and organic matter content of the sediment (both of which influence lead concentrations). In some areas, such as sites located downwind of smelters and tailings piles, the atmospheric deposition of lead might also complicate vertical trends in lead concentrations. Various methods have been developed to reduce those uncertainties, such as analyzing multiple floodplain cores at a site and normalizing metal concentrations on the basis of a specific grain size, organic matter content, or conservative element[3] (Horowitz 1991; Harrison et al. 2003; Bonotto and de Lima 2006; Miller and Orbock Miller 2007).

In addition to analyzing floodplain deposits, the approach has often been applied to sediments in the bed of a reservoir (Von Gunten et al. 1997; Kähkönen et al. 1998; Müller et al. 2000; Audry et al. 2004; Miller and Mackin 2013). The disruption of stream flow and sediment transport by a dam typically results in higher rates of nearly continuous sediment accumulation in a reservoir (Müller et al. 2000; Arnason and Fletcher 2003) and makes it easier to create a high-resolution record of historical changes in sediment and sediment-associated metal loadings into a river. In addition, thicker depositional units reduce the potential influences of the postdeposition migration of lead in the deposits (Callender 2000; Audry et al. 2004). In other cases, other types of sediments might be examined to complement the historical sedimentary record (Box 4-3).

[3]A *conservative element* is one that is uniformly distributed throughout the sampled core. If the ratio of another element to a conservative element changes with depth in the core, the change is a signal that the other element is being depleted from or being supplied to the system.

Mineralogic and Morphologic Analyses of Bulk Samples

It is not uncommon for the geologic materials associated with a lead source to contain a unique set of miner-

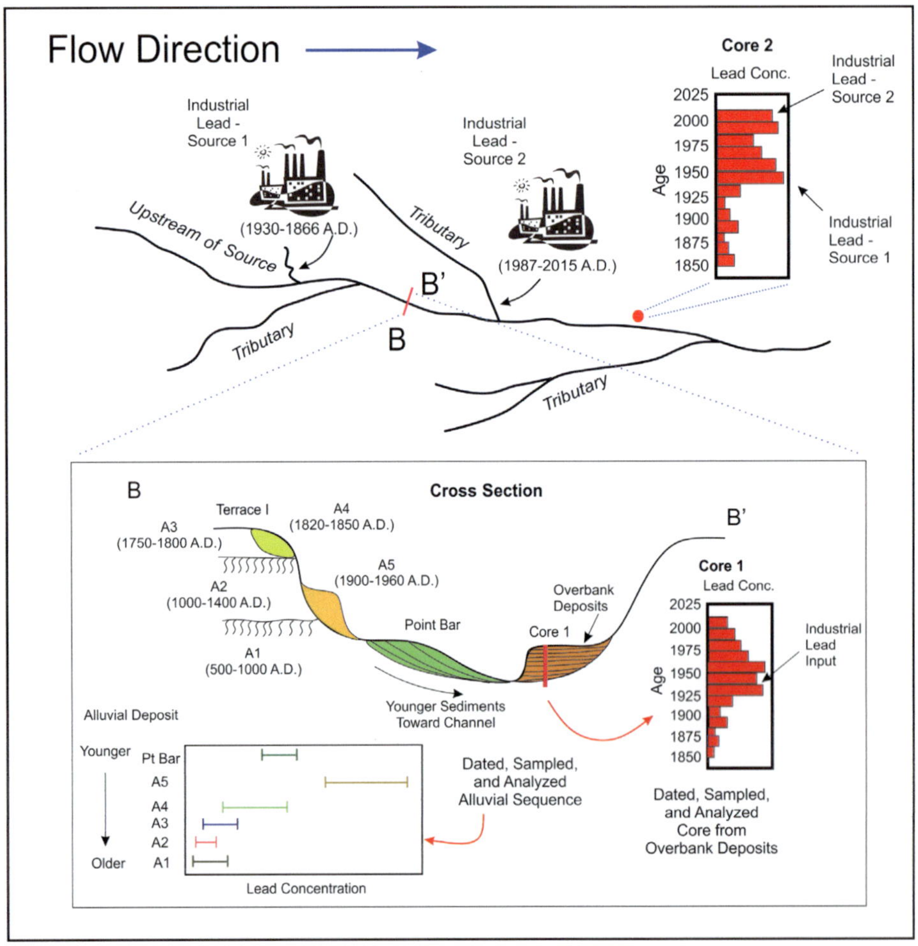

FIGURE 4-7 Schematic diagram illustrating how the age of a sedimentary deposit can be combined with lead-concentration data to determine the source of lead in river sediments. Sources are determined by linking temporal peaks in lead concentration to known anthropogenic inputs at the time of sediment deposition. Variations associated with natural inputs of lead can be assessed by examining prehistoric deposits (not shown).

als. Unique minerals or mineral assemblages have been used effectively to determine the origin of sediment from point and nonpoint (diffuse) sources along contaminated and uncontaminated rivers. Two of the most common analyses focus on clay minerals or heavy-mineral assemblages (Jeong and McDowell 2003; Eberl 2004; Gingele and DeDeckker 2005; Benedetti et al. 2006; Chen et al. 2009; Hardy et al. 2010; Bernárdez et al. 2012), although other mineral groups might also be useful depending on the nature of the source.

The effectiveness of the use of clay minerals stems from the fact that they reflect the composition and weathering characteristics of rock types that underlie the basin. Different rock types generally yield different clay minerals on weathering. Of importance to lead attribution in and around Superfund sites, the use of clay minerals has the potential to differentiate between sediments derived from nonmineralized rocks and mineralized or mined rock units. For example, differences in clay mineralogy were used by Jeong and McDowell (2003) to differentiate between mine tailings (enriched in the clay mineral chlorite), Ontonagon River sediments (with low chlorite and high illite), and a Wisconsin red-clay deposit (with low illite and high expandable-clay content) along the south-central coast of Lake Superior.

Heavy minerals, particularly sulfide minerals, have also been used to assess the origin of sediment and sediment-associated trace metals (including lead) in mining districts (Schreck et al. 2005; Bernárdez et al. 2012). Schreck et al. (2005), for example, found that minerals as-

BOX 4-3 Historical Analysis of Lake, Bog, and Coastal Sediments

In addition to the analysis of river sediments, sedimentary records from lakes, peat bogs, and coastal areas can provide valuable insights into the sources and magnitude of lead loadings into the environment. The analysis is similar to the methods applied to overbank floodplain and reservoir deposits in that the sediments are sampled, analyzed, and dated, and this allows the development of a record of the temporal variations in lead influx into the water body. The temporal record can then be compared with an understanding of the history of anthropogenic lead use in the region, including timeframe and extent of activity, changes in regulatory measures, and location of sources.

The input of lead and other trace metals into lakes occurs by a combination of wet and dry atmospheric deposition and the inflow of water and sediment from the surrounding drainage basin (Norton et al. 2004; Aebischer et al. 2015). It follows that the sedimentary deposits produced record changes in metal inputs from both atmospheric and basin sources. The analysis of lake sediments to determine atmospheric sources of lead deposition has been complemented by the analysis of peat bogs (Weiss et al. 1999, 2002; Bindler 2006; Kylander et al. 2009; Shotyk and Krachler 2010). Some peat bogs receive water, nutrients, and lead almost exclusively from the atmosphere; their sedimentary records are particularly useful because they are not substantially affected by either surface-water or groundwater inputs (Damman 1986; Shotyk 1988).

Assessment of lead sources in and adjacent to Superfund sites necessarily requires distinguishing local from global or regional atmospheric lead sources. Two approaches are commonly used to differentiate the two. First, multiple lakes are cored and analyzed; some receive runoff and lead from local sources, and others receive lead only from atmospheric sources. Norton et al. (2004), for example, analyzed the sedimentary records contained in three lakes in Massachusetts that were adjacent to a Superfund site. Two of the lakes received lead from their drainage basins. The third lake was in a relatively undisturbed basin and therefore received lead primarily from atmospheric sources. Comparison of data on the three lakes provided insights into the magnitude and timing of contamination of the environment by local sources.

The second, and more common, approach is to combine lead-concentration data with lead isotopic data (see, for example, Shirahata et al. 1980; Petit et al. 1984; Graney et al. 1995; Callender 2005; Aebischer et al. 2015). As presented earlier, comparison of the changes in the lead isotopic composition of the sediments can often be linked to suspected lead sources. A more recent, and potentially useful, approach is to analyze the cores for lead, lead isotopes, and other isotopes (such as zinc and iron isotopes) to assess source. For example, Aebischer et al. (2015) used lake sediments in Schefferville, Quebec, to assess the effect of mining on aquatic ecosystems. Combining lead-concentration data with lead, zinc, and iron isotopic data allowed the identification of multiple sediment sources of lead in the lake. A number of recent studies have focused on zinc isotopic records in lakes to assess the input of metals into the environment from smelting and other atmospheric sources (Sonke et al. 2002, 2008; Cloquet et al. 2006; Weiss et al. 2007; Mattielli et al. 2009; Thapalia et al. 2010).

sociated with the Kupferschiefer ores, which outcropped in the Mansfeld mining district of central Germany, were dominated by covellite, chalcopyrite, and sphalerite. The mineral assemblage differed from the waste materials that were associated with metalliferous flue dust and smelter wastes, which were dominated by slag particles; the differences allowed their contributions to the river to be determined. However, mineral assemblages were similar between mining wastes composed of low-grade ore and natural outcrops of the ore deposits. Determining the importance of the latter sources depended on spatial variations in the waste piles in the basin. A potential shortcoming associated with the use of heavy minerals is that weathering of the particles after deposition might alter the mineralogic characteristics of the source sediments.

Other minerals might also be used to assess sediment contributions from mining operations. In Southeast Missouri, mineral analyses showed that chat particles sampled from the Leadwood, Federal, and National tailings piles were composed almost exclusively of dolomite, whereas sediments sampled from selected control sites were composed of greater than 95% chert and feldspar. Pavlowsky et al. (2010) argued that the quantity of dolomite in the Big River was an indicator of the relative contribution of mine tailings to the river.

The structure of fine-grained sediment is often examined to help to identify minerals. In other cases, the shape of such fine-grained sediments might provide direct insights into sediment and sediment-associated lead sources. As noted above, particles emitted from smelters are often spherical; that knowledge combined with an understanding of their composition can determine their source.

The structure of larger particles in rivers is typically the product of their degree of weathering at their source, abrasion during transport, and modifications after deposition. In general, the farther a particle has been moved along the river, the more rounded (smooth) it is. Thus, particle shape provides information on transport distance and the potential source in the basin (de Boer and Crosby 1995; Cardona et al. 2005). In mining areas, the mechanical grinding and crushing processes associated with the production of mill tailings might create particles characterized by a broken, jagged appearance that can differ from that of sediment derived from natural sources in the basin and allow their identification.

Mineralogic and Morphologic Analysis of Particle Separates

Because of the processes involved in the erosion, transport, and deposition of sediment in river systems, it is possible that sand-sized or dense particles will be derived from a source different from that of the light and fine-grained particles, such as silt and clay (see, for example, Miller et al. 2013b). Thus, the identification of lead source by using mineralogic, chemical, or isotopic methods might be complicated by analyzing mixed population of particles that differ in size or density. It might be possible to alleviate the problem by collecting and analyzing separates of the bulk sediment, which are characterized by a narrow range of particle size or density.

Multielement Fingerprinting and Modeling

The use of multielement geochemical fingerprinting techniques has increased in recent years (Franks and Rowan 2000; Collins et al. 2012, 2013; Rowan et al. 2012; Miller et al. 2015). For example, multielement methods have been applied to suspended sediments (Walling and Woodward 1995; Collins et al. 1997, 2001; Massoudieh et al. 2013), riverbed sediments (Collins et al. 2013; Evrard et al. 2013), floodplain deposits (Collins et al. 2012), and sediments in lakes, reservoirs, and wetlands (Miller et al. 2005, 2013b; Pittam et al. 2009; Rowan et al. 2012). The increased use has been promoted by recent advances in analytic techniques, such as inductively coupled plasma–mass spectrometry, that allow the rapid analysis of large numbers of elements at a reduced cost and by the ability to determine the source of contemporary and historic sediments and sediment-associated lead.

Conceptually, it is assumed that as sediments (and their associated lead) are eroded and transported from a source or source area they are combined with sediments from all other sources in the basin to form a mixture. The geochemical composition of the mixture depends on the composition of the source materials and the relative proportions of sediment derived from each source, provided that the physical and chemical characteristics of the sediment are not altered during transport, as might occur by elemental leaching or changes in the sediment grain size. If the compositions of the source materials and the mixture in question are determined (measured), it is possible to estimate the relative proportions that each source contributed to the mixture (Box 4-1).

Single and multielement geochemical fingerprinting in rivers has been used most extensively to determine nonpoint sources of sediment and sediment-associated contaminants in a river, whose sources might be defined by soil type, underlying geologic units, land-use categories, or tributaries (Russell et al. 2001; Douglas et al. 2003; Miller et al. 2005, 2015; Collins et al. 2009). In mining areas, the approach might provide insights into the quantity of lead eroded from mineralized upland soils. Multielement fingerprinting has not often been applied to point sources of contamination, such as a tailings pile, but it has been shown to be effective when it has been applied (Bird et al. 2010a,b).

The application of statistically based mixing models to river sediments has been criticized because the models can generate vastly different results that have similar levels of uncertainty, particularly when contributions from a source approach 0 or 100 % (Collins et al. 2010; Rowan et al. 2012). Much of the uncertainty is associated with variability inherent in the physical and geochemical characteristics of the source materials and the river sediment. To reduce and quantify uncertainty inherent in the modeling results, recent studies have explored the use of an increasing suite of statistical approaches, including the use of Monte Carlo sampling frameworks and Bayesian methods (Franks and Rowan 2000; Small et al. 2002, 2004; Collins et al. 2010, 2012, 2013; Rowan et al. 2012; Cooper et al. 2014; Nosrati et al. 2014; Miller et al. 2015; Palazón et al. 2015). The analyses have led to a general agreement that current attempts to apportion sediment and sediment-associated metals that use inverse (unmixing) models should include a sensitivity analysis for the input parameters and a detailed assessment of model uncertainty.

Other Source-Associated Tracers

In addition to the methods described above, a number of other characteristics have been used effectively to attribute sediment and sediment-associated metals in river sediments. They include magnetic properties (Yu and Oldfield 1993; Maher et al. 2009; Armstrong et al. 2010; Guzmán et al. 2010), particle color (Croft and Pye 2004; Martinez-Carreras et al. 2010), chemical speciation (Macklin and Dowsett 1989; Shikazono et al. 2012), and isotopic analyses, such as analyses of isotopes of strontium and neodymium (Weldeab et al. 2002; Padoan et al. 2011; Rosenbauer et al. 2013), copper (El Azzi et al. 2013), mercury (Blum and Bergquist 2007; Foucher et al. 2009; Sonke et al. 2010; Mil-Homens et al. 2013; Blum et al. 2014) and zinc (Sivry et al. 2008; Chen et al. 2009). Although the latter isotopes have not been applied extensively to the problem of lead-source attribution, their application holds considerable promise. Elements that have unique isotopic characteristics as a result of smelting, such as cadmium and zinc, might be particularly useful.

Modeling of Upland Erosion, Sediment Yields, and Transport

A wide array of surface-water modeling routines of varied degrees of complexity have been developed and applied to rivers to understand and predict sediment and sediment-associated contaminant-dispersal processes from both point and nonpoint sources. In general, the models are not designed specifically to determine the source of lead in river basins, and their successful application often depends on the availability and quality of topographic, hydrologic, geologic, and other forms of data. Given those constraints, the wide array of models available, and the ability to combine different models for specific purposes, the committee chose to exclude them from its investigative strategy. Nonetheless, such models might provide important insights into the source of sediment-associated lead. For example, in addition to known point sources of lead in and around a Superfund site, diffuse sources of natural lead exist in a river basin. The occurrence of lead in mineralized soils, which are eroded and delivered to a river during runoff events, might be particularly important in Superfund sites associated with mining districts. It might therefore be important to assess the contributions of lead derived from diffuse upland sources by erosion and mobilization processes. That objective might be addressed partly through the application of spatially distributed models that determine the location and magnitude of upland erosion and the amount of sediment that leaves the basin mouth. The distributed models typically partition the drainage basin into small parcels, each characterized by its own conditions, such as soil type, slope, and vegetation. Those data can then be used to assess soil mobilization and delivery to the river from specific parcel types, such as a specific type of soil or bedrock unit.

GROUNDWATER AND SURFACE WATERS

As noted in Chapter 3, concentrations of dissolved lead are typically low in surface waters or groundwater with a pH of 6-8. Thus, concentrations of dissolved lead tend to be relatively low in areas characterized by carbonate rocks (limestone and dolomite), such as in Southeast Missouri, that buffer pH near neutral and limit acid generation associated with the weathering of sulfide minerals, such as pyrite and galena, found in lead-bearing ores. The concentrations can be high in acidic water in other mining areas where limestone and other rock types that can neutralize acid water are generally not present. Here, the committee presents its investigative strategy for groundwater and surface waters (see Figure 4-8) and provides a brief discussion of the methods. The analysis of lead associated with water particles has been discussed in previous sections on soils and sediments.

Site Characterization and Hydrologic Conceptual Model

A first step in determining lead sources in water is to identify and characterize possible sources in and adjacent to the site. For example, if the focus is on river water, potential sources upstream of the site should be identified with potential on-site sources. Once possible sources are located, a conceptual model of the potential dispersal pathways via surface waters and groundwater can be developed by using existing data and standard hydrologic and hydrogeologic approaches. The conceptual model should address questions whose answers might depend on whether lead was found in groundwater from a well, in surface water from a river, stream, lake, or pond, or in both. Possible questions to consider include, What surface-water bodies are downgradient of the potential lead sources and thus might receive dissolved lead from the source? Do the adjacent streams receive groundwater that originates from the vicinity of the lead source? Is river water derived primarily from surface-water runoff, tributaries, or groundwater, and do these contain lead sources? What geologic materials come into contact with the groundwater and its flow path, and how do they influence flow (for example, limestone vs sandstone)? What are the depth to groundwater and the slope of the groundwater surface, and what are the sources of groundwater recharge? What are the vertical and lateral directions and rates of groundwater flow? Are there important anthropogenic sources of lead, such as lead-containing paint or plumbing, at the site that might dissolve and contribute to lead in water?

Once the conceptual model has been developed, it should be possible to define which sources can contribute lead to groundwater and surface waters. In some cases, such a hydrogeologic assessment might lead to a confident conclusion about the source of the dissolved lead that obviates further study. For example, a suspected lead source might be downgradient of a water-sampling location and be eliminated as a possible source, whereas an obvious upgradient source might be implicated if other lead sources are eliminated.

Groundwater and Surface-Water Sampling

If the conceptual hydrogeologic analysis does not elucidate the source with sufficient confidence, additional investigations are necessary. The sampling of groundwater should reflect the directions, depths, and rates of flow relative to the suspected lead sources and use accepted sampling protocols (see, for example, USGS 2006). Wells selected for sampling should accurately reflect the water chemistry of the hydrogeologic system under investigation and yield samples suitable for water analysis. At a given site, a sufficient number of appropriate wells for sampling or sufficient historical sampling data might not be available. Various factors can influence the integrity of a groundwater sample, and such characteristics as well

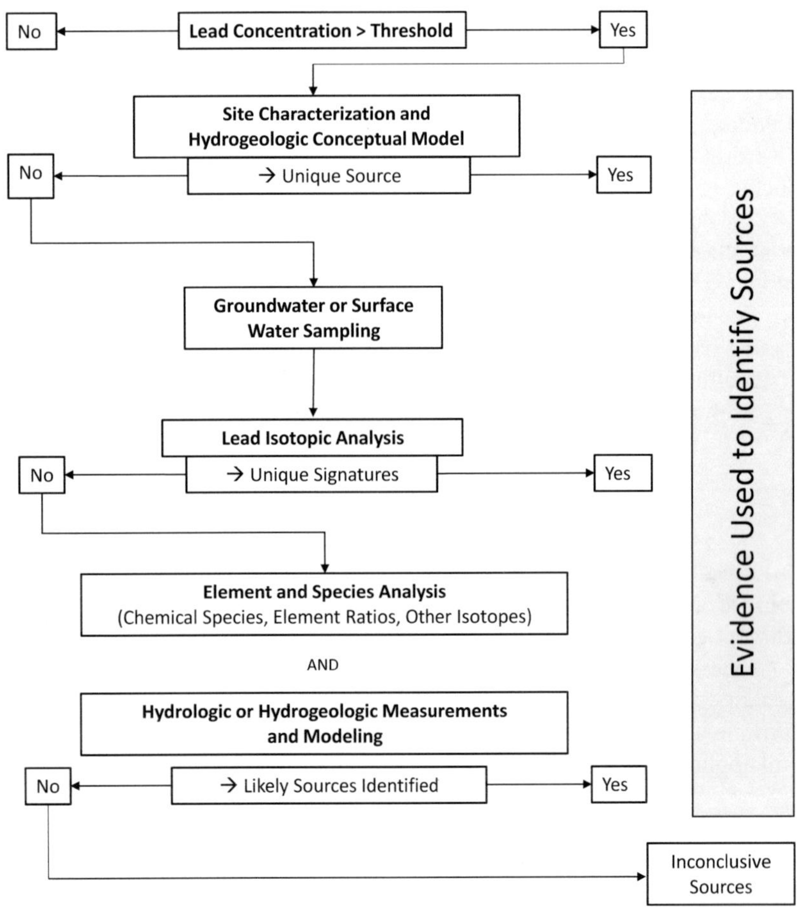

FIGURE 4-8 Example of an investigative strategy for attributing sources of lead in water at a Superfund site. The optimal sequence will depend on site-specific conditions, availability of analytic techniques, and cost and time considerations.

type, depth to groundwater, water-sampling depth based on the location of the well screen, pumping rate, water yield, temperature, pressure, and amount of suspended particulate matter (PM) should be considered in the sampling strategy. Wells that are not frequently pumped, such as monitoring wells, should be purged of standing water by pumping before sampling. If dissolved concentrations are of interest, water samples are filtered through a 0.45-μm pore-size membrane and analyzed for lead and other constituents. If total concentrations are desired, unfiltered and acid-digested samples are analyzed. Given the generally low lead concentrations in surface waters and groundwater, all sampling and filtration equipment should be lead-free. All water-sample handling, preservation methods, and storage before analysis should follow standard protocols for trace metals to avoid changes after sampling that might affect measured concentrations (EPA 1996a; Lurry and Kolbe 2000).

In rivers, two approaches to sampling are often used to characterize concentrations and sources of dissolved lead. First, samples are collected at multiple times at fixed locations to characterize temporal variations in lead inputs into the river, particularly during changes in flow or flood events. A monitoring station established along a river has the potential to collect multiple samples as water levels rise and fall during a flood event. Second, one or more synoptic sampling campaigns might be conducted along river segments of interest; synoptic sampling is sampling of a river at many locations at the same time. The samples are obtained from the main channel and all known inflow points—such as tributaries, culverts, irrigation ditches, and urban sewer systems—along the river during a relatively short period of stable flow. River flow (discharge) is also measured so that the amount of lead (mass/time) that passes any point along the river can be calculated from the product of lead concentration (mass/m^3) in the river and flow (m^3/time). Thus, the analysis provides a snapshot of the variations in elemental concentrations and load along the river that can then be interpreted by using mass balances and inferred or observed lead inputs.

Methods for sampling standing bodies of water, such as lakes and ponds, are generally similar to those used for rivers although strategies for the timing and location of sample collection might differ. The number of samples to collect, the distribution of sampling locations, and the interval between sampling events should be determined on the basis of the size, depth, inflow and outflow points, water residence time, seasonal variations, and study objectives for the particular water body. For example, samples collected at the inflows and outflows of the lake or pond might be compared with representative samples collected from the interior. Sampling as a function of depth in different locations might provide information about water mixing or settling of PM that might affect concentrations. Depth sampling is usually done with a sampling device that is dropped on a weight to a desired depth and opened or closed with an activator. If it is important to assess changes in water chemistry as a function of time, time between samplings of lakes or ponds might be much longer than times between samplings of rivers and will depend on the residence time of the water and possibly on seasonal variations.

Lead Isotopic Analysis

As discussed above, lead isotopic abundances have been used extensively to determine the potential sources of lead in a wide array of media, including water. Specifically, a quantitative comparison is conducted between the isotopic signature of lead in water and possible sources, such as ores, tailings, nonmineralized rocks, and other lead-containing materials in the area. Some lead isotopic data might be available in the published literature. As noted, the method requires that the isotopic signatures of the possible lead sources be sufficiently different to warrant measurement of lead isotopes in water samples. That is, if all the possible sources of lead have the same isotopic signature, an expensive time-intensive isotopic analysis might not be appropriate (Aberg 2001; Komárek et al. 2008; Norrström and Knutsson 2012). If the known isotopic signatures of sources show that isotopic analysis might be useful, an unfiltered water sample could be digested for analysis of lead concentration and isotope ratios to determine the sources of lead in the water. The investigation might include application of a mixing model or other computational analysis. There might not be a unique isotopic signature, but the analytic results could add to the body of evidence and help to eliminate some sources. A total lead measurement includes dissolved lead and lead associated with suspended particles (Graham et al. 2006). A more in-depth analysis of lead isotopes in a filtered water sample (0.45 μm or smaller filter pore size) vs the particles retained on the filter might provide additional source information if the two fractions differ in their lead isotopic signatures.

Element and Species Analysis

If the isotopic signatures of the total, dissolved, or particulate lead do not yield a unique match to one or more of the lead sources or if more data are desired to build confidence in the lead-source attribution, further analysis of constituents in water will need to be undertaken. A first-order assumption is that high lead concentrations will occur at or immediately adjacent to a source and decrease with distance from the source as a result of dilution and a variety of attenuation processes. As noted in the soil and sediment sections, spatial variations in element concentrations and concentration ratios can provide qualitative insights into lead sources by identifying characteristic associations or ratios. Single-element and multielement analyses and measurement of dissolved chemical species and water-quality characteristics can be undertaken on surface waters and groundwater. For example, total, dissolved, and particulate samples could be analyzed for major, minor, and trace elements; for species, such as sulfate or phosphate; and for total dissolved carbon and organic carbon content. Standard measurements of water include temperature, conductivity, pH, dissolved oxygen, oxidation–reduction potential, and alkalinity (acid-neutralizing capacity) or acidity (base-neutralizing capacity). Stable isotopes of water (^{18}O, ^{16}O, ^{1}H, and ^{2}H), dissolved noble gases (^{4}He, ^{20}Ne, and ^{40}Ar), tritium (^{3}H), sulfur hexafluoride (SF_6), and radiocarbon (^{14}C) are among the tracers used to determine groundwater sources, flow paths, and water age (Clark and Fritz 1997; Cook and Bohlke 2000). Measurements of elements, species, isotopes, and water characteristics help to bound geochemical and hydrologic models that can predict water sources, flow paths, mixing, and reactions.

The chemical composition of particulate samples can be determined directly as discussed above or calculated on the basis of the difference between the measured total concentration and the dissolved concentration. Element ratios and element associations can then be compared with possible sources. Lead that is associated with particulate organic matter, for example, might point to an organic-rich water-body source. If filtered particles can be recovered, the particle size, structure, mineralogy, and composition could be investigated with SEM, XRD, or other techniques discussed above to provide insight into suspended-particle sources.

Hydrologic and Hydrogeologic Measurements and Modeling

Modeling of River Samples

Analysis of synoptic samples and temporal variations in lead concentrations can help to identify sources along rivers (Kimball 1997; Boughton 2001; Nimick and Cleasby 2001; Wirt et al. 2001; Walton-Day et al. 2005; Kimball et al. 2002, 2005, 2006, 2007, 2010; Church et al. 2007; Kimball and Runkel 2009). In most cases, the samples are obtained from the river channel and all known inflow points—such as tributaries, culverts, irrigation ditches, and urban sewer systems—during relatively short intervals of stable flows, which are also measured. The analysis provides a snapshot of the variations in elemental concentrations and loads along the river that reflect points of substantive elemental (lead) inputs. Instantaneous load is calculated as the product of constituent (lead) concentrations and river discharge at the sample point at the time of collection. Downstream variations in the loads in the river channel are used to determine the importance of each inflow point as a lead source and to determine the attenuation of lead along the river. Sources are identified by using a mass-balance approach that compares the cumulative load within the river channel with the cumulative load that enters the channel from upstream inflows (see, for example, Kimball et al. 2002). In many cases, the approach is combined with the use of elemental tracers, such as lead isotopes, to quantify more fully the contributions of a pollutant from a specific source.

An advantage of the synoptic approach is that the location and importance of unsampled, diffuse sources of lead from groundwater, which are often poorly documented, can be determined. A disadvantage of synoptic sampling is that it is difficult to apply during some periods of high river flow, but methods are being developed to overcome such issues (see, for example, Kimball et al. 2010). In some cases, the data collected during synoptic sampling can be used for in-stream modeling of solute transport, such as that of dissolved lead (Gooseff et al. 2008). An understanding of temporal variations in lead concentration at a site during a flood event can provide insights into the primary sources of contaminants (McDiffett et al. 1989; Evans and Davies 1998; House and Warwick 1998; Lloyd et al. 2016).

Geochemical, Hydrologic, and Hydrogeologic Models

Developing an overall assessment of sources of lead in water at a given site requires integration of multiple lines of evidence from isotopic and chemical measurements with hydrologic or hydrogeologic data. The data are typically analyzed and evaluated by using various statistical, numerical, and modeling approaches of varied complexity. Multivariate statistical methods, such as principal-components analysis or factor analysis, might be used to identify correlations or characteristic patterns among elements or species. Mixing models use differences in chemical or isotopic composition among the potential sources to determine whether the composition of a specific water can be represented by a mixture of the potential water-source compositions. Inverse mass-balance models compare differences in mass composition to deduce whether a specific water was derived from another along a flow path in contact with minerals and gases (Zhu and Anderson 2002). That model approach assumes a chemical steady state (no net change in chemical species) and is based strictly on mass differences and mass transfer between water and the minerals and gases that are assumed to be in contact with it. Mass-balance models also typically ignore hydrodynamic dispersion or diffusion processes that might affect solute concentrations (Zhu and Anderson 2002). Although potentially informative, mass-balance models consider net chemical mass differences and do not produce a unique solution; that is, multiple combinations of water types, minerals, and gases might fit the water composition in question and depend on the water, mineral, and gas phases assumed in the model.

Models that rely solely on mass differences are distinguished from chemical-speciation models that rely on mass balances and thermodynamics to establish the dominant chemical reactions and chemical species that are present. On the basis of thermodynamic equilibrium constants and kinetic parameter estimates, chemical-speciation models are able to calculate the dominant aqueous species and complexes, the amount of dissolution and precipitation of minerals, the solubility of gases in water, and the extent of surface sorption reactions. Several programs are available and include free packages, such as PHREEQC (Parkhurst and Appelo 1999) and MINTEQA2 (EPA 1999a). Chemical equilibrium calculations, particularly ones that include sorption reactions, can help to elucidate possible chemical reactions that lead to mobilization of lead from sources or how lead might be attenuated along a flow path. For example, streams contaminated by lead, zinc, and cadmium from mine waste runoff and mine-drainage discharge in the Tri-State district in Oklahoma, Missouri, and Kansas were studied with a combination of runoff and discharge measurements by using a conservative tracer, statistical analyses, mass-balance calculations, and equilibrium chemical modeling to evaluate lead sources, metal fluxes, and chemical reactions that mobilize or attenuate lead downstream of waste piles and mine-drainage input (Carroll et al. 1998; Schaider et al. 2014).

A variety of water-quality modeling routines might be applied to rivers, such as the Water Quality Assessment Simulation (Ambrose et al. 1988), which includes methods to model advection, dispersion, and point and diffuse mass loading processes and can be linked to hydrodynamic and sediment transport models to assess flow and particulate transport. For groundwater, hydrologic gradients, pump tests to calculate aquifer storage and characteristics, tracer studies to map flow paths and determine residence time, and geologic and hydraulic information are used to support aquifer and groundwater basin models. A widely used package is MODFLOW, the US Geological Survey three-dimensional finite-difference groundwater model (Harbaugh 2005). Originally developed solely as a groundwater-flow simulation code, MODFLOW has been greatly expanded to include simulation of coupled groundwater and "surface-water systems, solute transport, variable-density flow (including saltwater), aquifer-system compaction and land subsidence, parameter estimation, and groundwater management" (Brovelli and Moreno 2014). Surface-water or groundwater models are increasingly used in conjunction with analysis and modeling of chemical and isotopic data to provide a comprehensive evaluation of contaminant sources, transport, and transformation at a study site (Appelo and Postma 2005; Bedekar et al. 2016).

AIR

Like the investigative approaches presented for soils, sediments, and water, attribution of sources of airborne lead typically requires field assessment, laboratory analysis, and modeling. The laboratory analysis of lead-containing PM that might be collected by air monitors is similar to the laboratory analysis of soils described above. Therefore, the committee focuses here on field assessment and modeling approaches that are used to distinguish sources of airborne lead. The committee's investigative strategy is a tiered approach illustrated in Figure 4-9; each tier increases in complexity to achieve source attribution. Tier I begins by conducting chemical speciation on archived filter remnants after lead analysis has been conducted, Tier II includes field sampling to develop site-specific source signatures for input into receptor modeling, and Tier III includes more complete ambient and source sampling to provide additional evidence via a combination of source-dispersion and receptor modeling. Although the focus here is on traditional particle analysis, other approaches are possible. As noted above, Prapaipong et al. (2008) used oak trees to determine dispersion of lead emissions from a smelter, and other researchers have used lichen, rainfall, and snowpack sampling to understand spatial gradients of emissions from smelters and other sources (see, for example, Carignan and Gariepy 1995; Simonetti et al. 2000; Spiro et al. 2004; Telmer et al. 2004; Aznar et al. 2008; Graney et al. 2012; Graney and Landis 2013; Sherman et al. 2015).

Two types of models are used in the committee's investigative strategy. First, receptor models start with pollutant concentrations measured at receptors and attempt to identify and quantify contributions of different sources on the basis of source characteristics (Watson et al. 2016). Specifically, receptor models analyze patterns of chemical composition, particle size, temporal variability, and spatial gradients that are related to specific sources and take into consideration measurement uncertainty and source variability to determine contributions from specific sources that reach the receptors (Watson et al. 2002; Watson and Chow 2005, 2013, 2015; Watson et al. 2016). Second, source-dispersion models start with emission estimates related to potential sources—such as surface soils, road dust, and mine tailings—and calculate meteorologic transport of pollutants to receptors. The receptors are usually neighborhood, urban, or regional air-quality monitoring sites that are established to determine compliance with ambient air-quality standards. The US Environmental Protection Agency (EPA) maintains the Web site Support Center for Regulatory Atmospheric Modeling (SCRAM), which describes EPA-accepted air-quality models in great detail and provides free downloads for the accepted model software.[4]

[4]See https://www.epa.gov/scram.

Tier I: Analyze Composition of Archived Filters

The first tier in the committee's investigative strategy involves further analysis of samples that have already been collected from air-monitoring stations. As noted in Chapter 3, high airborne lead concentrations in total suspended particles (TSP) at monitoring sites of the Missouri Department of Natural Resources (see Figure 3-6) indicate the presence of lead-contaminated resuspended dust in the Southeast Missouri Lead Mining District (MO DNR 2016). Archived filter-sample remnants for which 24-hour TSP lead concentrations exceeded 0.15 $\mu g/m^3$ could be submitted for PM composition analyses, which could include multielement analysis (Watson et al. 1999) with XRF, inductively coupled plasma–atomic emission spectroscopy, and inductively coupled plasma–mass spectrometry; analysis of anions (such as chloride, nitrate, and sulfate) and cations (such as water-soluble sodium, magnesium, potassium, calcium, and ammonium) with ion chromatography; and thermal-optical carbon analysis of organic and elemental-carbon and thermal-carbon fractions. Lead concentrations should be compared with those reported by the Missouri Department of Natural Resources for quality assurance. Material balances of measured species should be performed to ensure that all major PM components are measured, that is, that the sum of the measured chemical species adequately explains the gravimetric mass (Chow et al. 2015).

High ambient PM lead concentrations might have various sources. Watson et al. (2016) recommend that additional indicators be obtained from archived filter-sample remnants to reduce source apportionment uncertainties. Such indicators include commonly found mineral-related elements (aluminum, silicon, potassium, calcium, and iron), elements related to mining activities (such as lead, zinc, copper, arsenic, and silver), stable lead isotopes (^{204}Pb, ^{206}Pb, ^{207}Pb, and ^{208}Pb) that can be analyzed after hot-block acid digestion procedures (with a combination of nitric, hydrochloride, and hydrofluoric acids) established by EPA for lead analysis with inductively coupled plasma-mass spectrometry (EPA 1996b, 1999b). Contaminated fugitive dust might be the main contributor to atmospheric lead concentrations, and additional measurements—such as the carbonate carbon and rare earth elements (such as cerium, dysprosium, erbium, europium, gadolinium, holmium, lanthanum, lutetium, neodymium, praseodymium, samarium, terbium, thulium, and ytterbium) and particle structure—can also be considered.

Descriptive data analysis can be conducted to understand PM composition and to examine temporal and spatial variations. Comparison of similarities and differences between lead-isotope ratios and abundances of rare earth elements among sites serves as the starting point for examining site-to-site and month-to-month variations (Felix et al. 2015). Because chemical composition varies with geologic source as illustrated in Figure 4-10, that information can be used in source attribution to define source categories. Receptor modeling of about 200–300 samples with the Effective Variance Positive Matrix Factorization and UNMIX solutions to the chemical-mass balance equations can be conducted to identify and quantify major source contributions of PM and lead (EPA 2016).

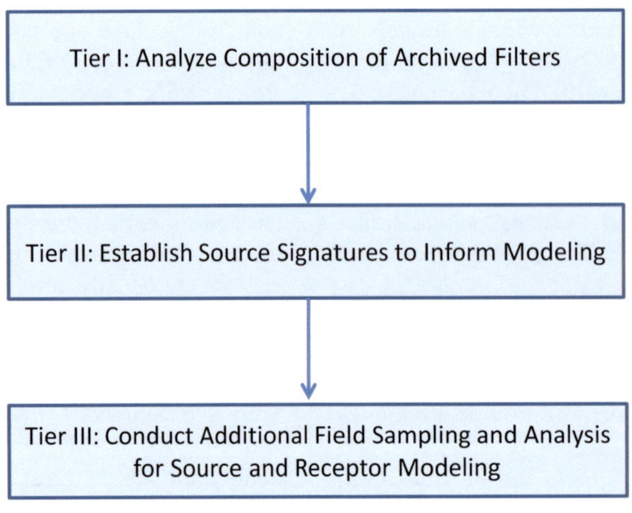

FIGURE 4-9 Investigative strategy for identifying sources of lead in air. The progression from one tier to the next represents increasing complexity.

Preliminary source contributions can serve as a guide in determining the type of source signatures that need to be acquired in Tier II and the extent of field sampling needed for Tier III.

Tier II: Establish Source Signatures for Receptor Model Input

Figure 4-11 illustrates the strategy for establishing source signatures. Source signatures of major source types—such as vehicle exhaust, aircraft fuel combustion, lead smelting, and glass manufacturing—from the EPA source-profile library (SPECIATE; EPA 2017a) can be compiled with the available local source signatures to evaluate the adequacy of signatures for receptor modeling. Geological materials are often a major source of PM mass, so site-specific source signatures probably need to be defined. Surface-soil and paved-road dust surrounding sampling sites that are not in compliance with air-quality standards for lead should be collected, processed, and resuspended in a laboratory resuspension chamber into TSP, PM_{10}, $PM_{2.5}$, and PM_1 size fractions by using multiple filter substrates that will accommodate the same chemical speciation as that of archived filter-sample remnants for Tier I identification. For residential areas that have high lead concentrations, soil samples and house lead-based paint chips should also be collected, processed, and resuspended. Additional bulk-material sampling and analysis of storage piles, tailings piles, and mine haul roads that are near sites that have high lead concentration should also be considered. The source signatures identified here in Tier II serve as model input and verification for receptor modeling and as the basis of development of a speciated PM emission inventory for source-dispersion modeling (EPA 2016).

Tier III: Conduct Additional Field Sampling and Analysis for Source-Dispersion and Receptor Modeling

EPA's National Emissions Inventory (EPA 2017b) shows that lead from mining constitutes only 10% of Missouri emissions, but processes or activities that result in resuspended dust emissions are not well represented in current emission factors (EPA 1998). Because source-dispersion models use emission factors to translate emissions into ambient concentrations, the modeling results can be variable. To improve the accuracy of receptor and source-

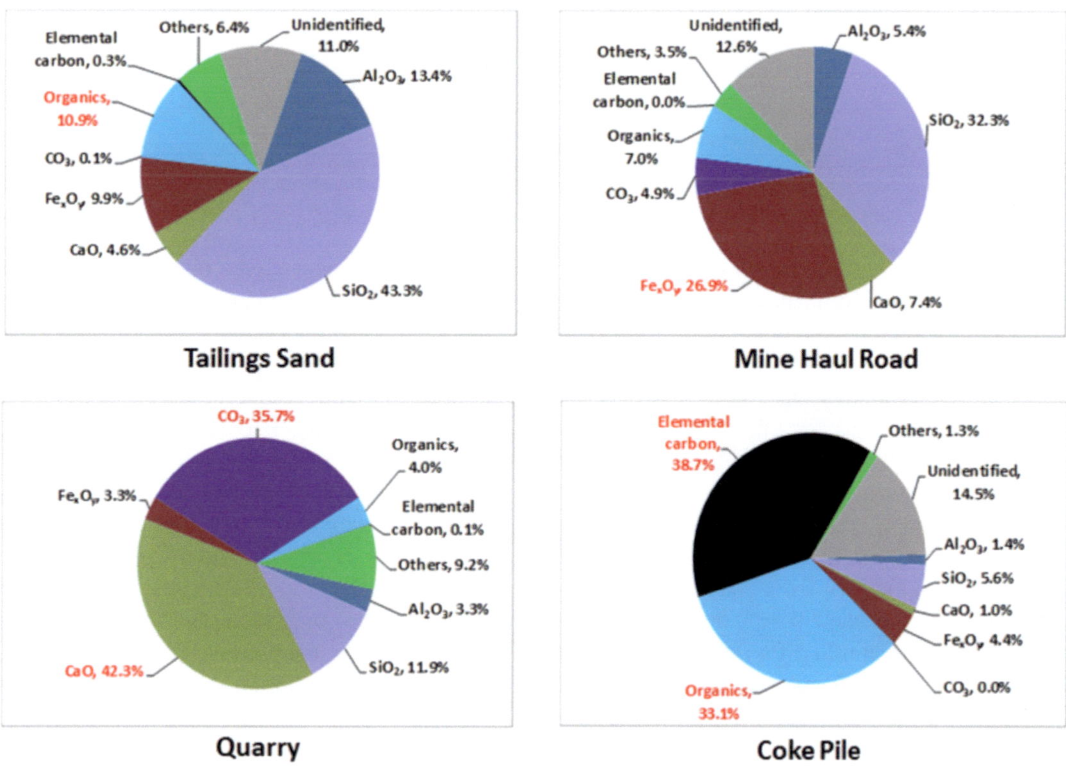

FIGURE 4-10 Chemical compositions differ among dust sources in the Canadian Oil Sands Regions. Source: Watson et al. (2014a,b). Reprinted with permission; copyright 2014, Desert Research Institute.

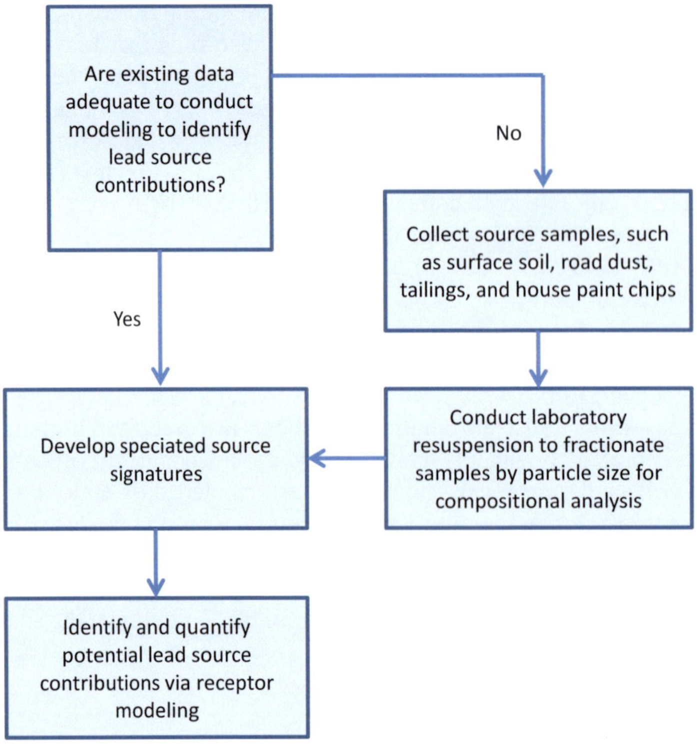

FIGURE 4-11 Expansion of the committee's Tier II strategy for establishing source signatures to inform receptor modeling that can help to identify and quantify potential lead sources.

dispersion models, field sampling is necessary to provide site-specific inputs. Thus, the committee's strategy for Tier III that is illustrated in Figure 4-12 includes varied field sampling to improve model inputs. For source-dispersion models, sensors that are mounted on-board a moving vehicle can be used to determine mechanical suspension and surface loadings (Kuhns et al. 2001), and portable wind tunnels can be used to measure wind erosion potential in various PM size fractions (Etyemezian et al. 2007; Sweeney et al. 2008; Wang et al. 2015). Source-dispersion and receptor models can both benefit from use of site-specific meteorologic variables, such as wind speed and direction, temperature, and relative humidity. Particle-size distributions illustrated in Figure 4-13 vary by geologic source categories and can be obtained by using an optical particle counter that provides particle-number concentrations in several size ranges, such as 0.01–10 µm. Finally, particle mass and chemical composition of lead and other geologically abundant elements (such as aluminum, silicon, potassium, calcium, iron, zinc, and titanium) can be determined by using a cascade impactor.

As indicated in Figure 4-12, model reconciliation can be used to identify uncertainties and limitations in receptor and source-dispersion modeling results and to improve model input datasets (EPA 2007). The goal here is to establish how well the receptor concentrations predicted by a source-dispersion model agree with the measurements collected at receptors and to assess whether receptor-model output compares favorably with the available information on source locations and signatures. Another approach is to use source-dispersion models in a parametric mode to predict chemical patterns at a receptor and then to use receptor models to identify the patterns that best reproduce the ambient measurements (Watson and Chow 2014).

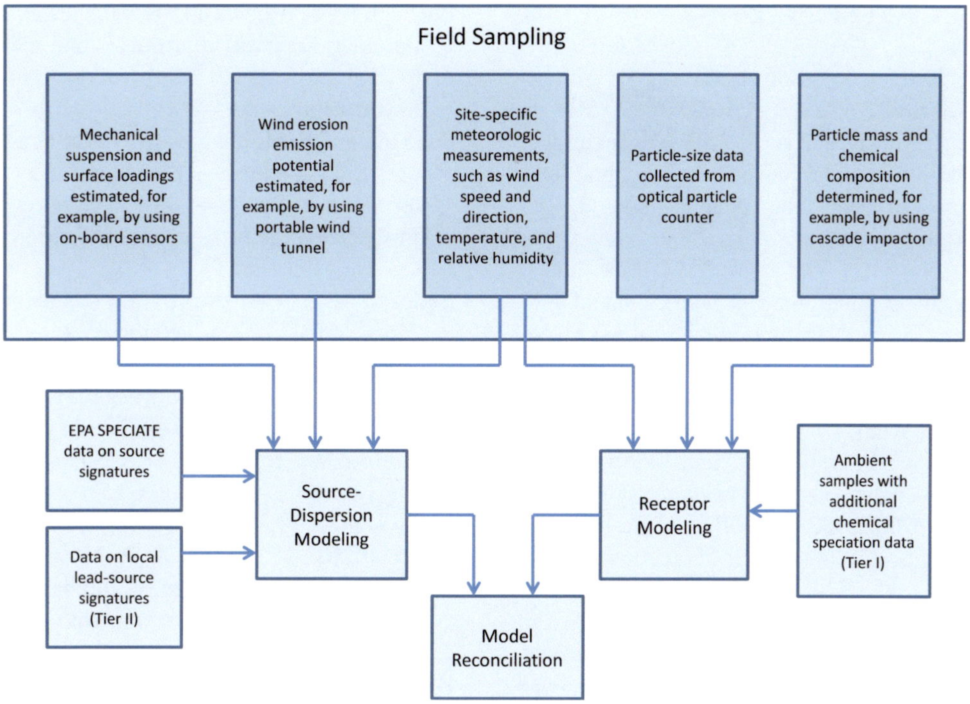

FIGURE 4-12 Expansion of the committee's Tier III strategy for estimating emission potential and particle-size distribution to provide inputs to modeling for source attribution.

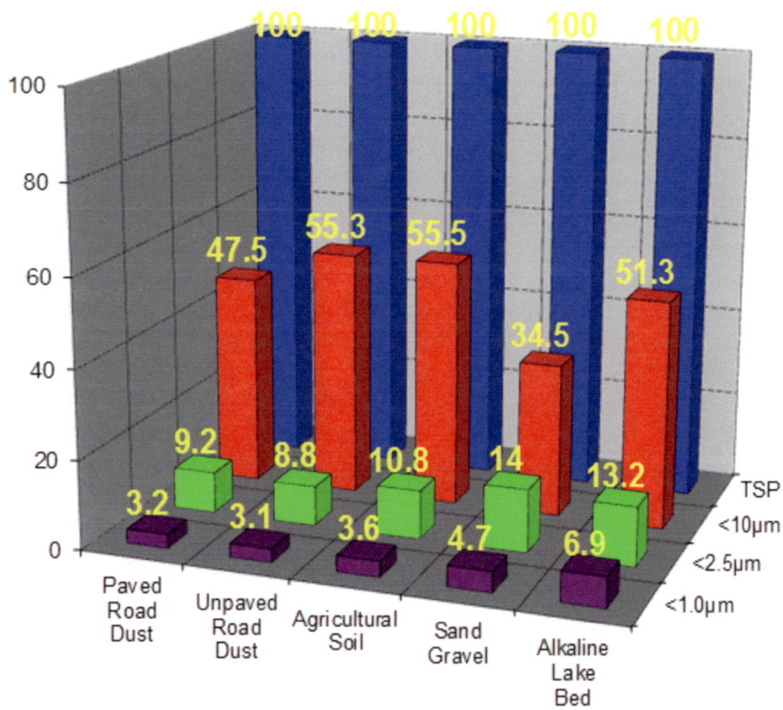

FIGURE 4-13 Comparison of size distribution of particles from Fresno (CA) paved-road dust, Bakersfield (CA) unpaved-road dust, Stockton (CA) agricultural soil, Visalia (CA) sand and gravel, and Owens Lake (CA) alkaline lake-bed sediments. Vertical axis and numbers above bars are percentages. Source: Chow et al. (1994). Reprinted with permission; copyright 1994, *Atmospheric Environment*.

REFERENCES

Åberg, G.E. 2001. Tracing pollution and its sources with isotopes. Water Air Soil Pollut. 130(1):1577-1582.

Abu-Zeid, M., A. Baghdady, and H. El-Etr. 2001.Textural attributes, mineralogy and provenance of sand dune fields in the greater Al Ain area, United Arab Emirates. J. Arid. Environ. 48(4):475-499.

Aebischer, S., C. Cloquet, J. Carignan, C. Maurice, and R. Pientiz. 2015. Disruption of the geochemical metal cycle during mining: Multiple isotope studies of lake sediments from Schefferville, subarctic Québec. Chem. Geol. 412:167-178.

Ambrose, R.B., T.A. Wool, J.P. Connolly, and R.W. Schanz. 1988. WASP4, A Hydrodynamic and Water Quality Model - Model Theory, User's Manual, and Programmer's Guide. EPA/600/3-87-039. Office of Research and Development, U.S. Environmental Protection Agency, Washington, DC.

Anderson, S.P., W.E. Dietrich, and G.H. Brimhall. 2002. Weathering profiles, mass balance analysis, and rates of solute loss: Linkages between weathering and erosion in a small, steep catchment. Geol. Soc. Am. Bul. 114(9):1143-1158.

Appelo, C.A.J., and D. Postma. 2005. Geochemistry, Groundwater and Pollution, 2nd Ed. Leiden, The Netherlands: A.A. Balkema.

Armstrong, A., B. Maher, and J. Quinton. 2010. Enhancing the magnetism of soil: The answer to soil tracing? P. 2805 in EGU General Assembly 2010, May 2-7, 2010, Vienna, Austria.

Arnason, J.G., and B.A. Fletcher. 2003. A 40+ year record of Cd, Hg, Pb, and U deposition in sediments of Patroon Reservoir, Albany County, NY, USA. Environ. Pollut. 123(3):383-391.

Arribas, J., S. Critelli, E. Le Pera, and A. Tortosa. 2000. Composition of modem stream sand derived from a mixture of sedimentary and metamorphic source rocks (Henares River, Central Spain). Sediment Geol. 133:27-48.

Audry, S., J. Schäfer, G. Blanc, and J.M. Jouanneau. 2004. Fifty-year sedimentary record of heavy metal pollution (Cd, Zn, Cu, Pb) in the Lot River reservoirs (France). Environ. Pollut. 132(3):413-426.

Ayrault, S., M. Roy-Barman, M.F. Le Cloarec, C. Rianti Priadi, P. Bonté, and C. Göpel. 2012. Lead contamination of the Seine River, France: Geochemical implications of a historical perspective. Chemosphere 87(8):902-910.

Aznar, J.C., M. Richer-Lafleche, and D. Cluis. 2008. Metal contamination in the lichen Alectoria sarmentosa near the copper smelter of Murdochville, Quebec. Environ. Pollut 156(1):76-81.

Bábek, O., M. Faměra, K. Hilscherová, J. Kalvoda, P. Dobrovolný, J. Sedláček, J. Machát, and I. Holoubek. 2011. Geochemical traces of flood layers in the fluvial sedimentary archive: Implications for contamination history analyses. Catena 87(2):281-290.

Balcaen, L., L. Moens, and F. Vanhaecke. 2010. Determination of isotope ratios of metals and metalloids by means of ICP mass spectrometry for provenanceing purposes. Spectrochim. Acta Part B At. Spectrosc. 65(9):769-786.

Basu, A., and E. Molinaroli. 1991. Reliability and application of detrital opaque Fe-Ti oxide minerals in provenance determination. Pp. 55-65 in Developments in Sedimentary Provenance Study. Special Publication Vol. 57. London: Geological Society.

Bedekar, V., E.D. Morway, C.D. Langevin, and M. Tonkin. 2016. MT3D-USGS Version 1: A U.S. Geological Survey Release of MT3DMS Updated with New and Expanded Transport Capabilities for Use with MODFLOW. U.S. Geological Survey Techniques and Methods 6-A53. Reston, VA: U.S. Geological Survey [online]. Available: https://pubs.usgs.gov/tm/06/a53/tm06a53.pdf [accessed September 21, 2017].

Benedetti, M., M. Raber, M. Smith, and L. Leonard. 2006. Mineralogical indicators of alluvial sediment sources in the Cape Fear River basin, North Carolina. Phys. Geogr. 27(3):258-281.

Bernárdez, P., R. Prego, S. Giralt, J. Esteve, M. Caetano, S. Parra, and G. Francés. 2012. Geochemical and mineralogical characterization of surficial sediments from the Northern Rias: Implications for sediment provenance and impact of the source rocks. Mar. Geol. 291-294:63-72.

Bindler, R. 2006. Mired in the past-looking to the future: Geochemistry of peat and the analysis of past environmental changes. Glob. Planet. Chang. 53(4):209-221.

Bird, G. 2011. Provenancing anthropogenic Pb within the fluvial environment: Developments and challenges in the use of Pb isotopes. Environ. Int. 37(4):802-819.

Bird, G., P.A. Brewer, M.G. Macklin, M. Nikolova, T. Kotsev, M. Mollov, and C. Swain. 2010a. Quantifying sediment-associated metal dispersal using Pb isotopes: Application of binary and multivariate mixing models at the catchment-scale. Environ. Pollut. 158(6):2158-2169.

Bird, G., P.A. Brewer, M.G. Macklin, M. Nikolova, T. Kotsev, M. Mollov, and C. Swain. 2010b. Pb isotope evidence for contaminant-metal dispersal in an international river system: The lower Danube catchment, Eastern Europe. Appl. Geochem. 25(7):1070-1084.

Blum, J.D., and B.A. Bergquist. 2007. Reporting of variations in the natural isotopic composition of mercury. Anal. Bioanal. Chem. 388(2):353-359.

Blum, J.D., L.S. Sherman, and M.W. Johnson. 2014. Mercury isotopes in earth and environmental science. Annu. Rev. Earth Planet. Sci. 42:249-269.

Bonotto, D., and J. de Lima. 2006. 210Pb-derived chronology in sediment cores evidencing the anthropogenic occupation history at Corumbatai river basin, Brazil. Environ. Geol. 50:595-611.

Boughton, G.K. 2001. Metal Loading in Soda Butte Creek Upstream of Yellowstone National Park, Montana and Wyoming: A Retrospective Analysis of Previous Research; and Quantification of Metal Loading, August 1999. Water Resources Investigation Report 01-4170. U. S. Geological Survey [online]. Available: https://pubs.usgs.gov/wri/wri014170/pdf/wri004170.pdf [accessed June 23, 2017].

Brantley, S.L., and M. Lebedeva. 2011. Learning to read the chemistry of regolith to understand the Critical Zone. Annu. Rev. Earth Planet. Sci. 39:387-416.

Brimhall, G., and W.E. Dietrich. 1987. Constitutive mass balance relations between chemical composition, volume, density, porosity, and strain in metasomatic hydrochemical systems: results on weathering and pedogenesis. Geochim. Cosmochim. Acta 51(3):567-587.

Brovelli, M., and R. Moreno. 2014. Free and Open Souce Software and Web Services Specializing in the Water Resources Domain. Webminar, September 2014 [online]. Available: http://www.asprs.org/a/webinarseries/Brovelli_Moreno_ASPRS_Webminar_Part1_9_26_2014.pdf [accessed June 23, 2017].

Burgess, M. 2016. National Academy of Sciences Sources of Lead Contamination at or Near Superfund Sites. Presentation at the First Meeting on Sources of Lead Contamination at or near Superfund Sites, November 21-22, 2016, Washington, DC.

Burt, R., M.A. Wilson, T.J. Keck, B.D. Dougherty, D.E. Strom, and J.A. Lindahl. 2003. Trace element speciation in selected smelter-contaminated soils in Anaconda and Deer Lodge Valley, Montana, USA. Adv. Environ. Res. 8(1):51-67.

Buseck, P.R., and M. Posfai. 1999. Airborne minerals and related aerosol particles: Effects on climate and the environment. Proc. Nat. Acad. Sci. USA 96(7):3372-3379.

Cabala, J., and J. Teper. 2007. Metalliferous constituents of rhizospere soils contaminated by Zn-Pb mining in southern Poland. Water Air Soil Pollut. 178(1):351-362.

Callender, E. 2000. Geochemical effects of rapid sedimentation in aquatic systems: Minimal diagenesis and the preservation of historical metal signatures. J. Paleolimnol. 23(3):243-260.

Callender, E. 2005. Heavy metals in the environment—Historical trends. Pp. 67-105 in Environmental Geochemistry, Treatise on Geochemistry Vol. 9, B. Sherwood Lollar, ed. Amsterdam: Elsevier.

Cardona, J.P.M., J.M.G. Mas, A.S. Bellon, S. Domiinguez-Bella, and J.M. Lopez. 2005. Surface textures of heavy mineral grains: A new contribution to provenance studies. Sediment Geol. 174(3-4):223-235.

Carignan, J., and C. Gariepy. 1995. Isotopic composition of epiphytic lichens as a tracer of the sources of atmospheric lead emissions in southern Quebec, Canada. Geochim Cosmochim. Acta 59(21):4427-4433.

Carroll, S.A., P.A. O'Day, and M. Piechowski. 1998. Rock-water interactions controlling zinc, cadmium, and lead concentrations in surface waters and sediments, U.S. Tri-State Mining District. 2. Geochemical interpretation. Environ. Sci. Technol. 32(7):956-965.

Chen, J.B., J. Gaillardet, P. Louvat, and S. Huon. 2009. Zn isotopes in the suspended load of the Seine River, France: Isotopic variations and source determination. Geochim. Cosmochim. Acta 73(14):4060-4076.

Cheng, H., and Y. Hu. 2010. Lead (Pb) isotopic fingerprinting and its applications in lead pollution studies in China: A review. Environ. Pollut. 158(5):1134-1146.

Chiaradia, M., B.E. Chenhall, A.M. Depers, B.L. Gulson, and B.G. Jones. 1997. Identification of historical lead sources in roof dusts and recent lake sediments from an industrialized area: Indications from lead isotopes. Sci. Total. Environ. 205(2-3):107-128.

Chow, J.C., J.G. Watson, J.E. Houck, L.C. Pritchett, C.F. Rogers, C.A. Frazier, R.T. Egami, and B.M. Ball. 1994. A laboratory resuspension chamber to measure fugitive dust size distributions and chemical compositions. Atmos. Environ. 28(21):3463-3481.

Chow, J.C., D.H. Lowenthal, L.W.A. Chen, X. Wang, and J.G. Watson. 2015. Mass reconstruction methods for PM2.5: A review. Air Qual. Atmos. Health 8(3):243-263.

Church, S.E., D.M. Unruh, D.L. Fey, and T.C. Sole. 2004. Trace Element and Lead Isotopes in Streambed Sediment in Streams Affected by Historical Mining. Chapter D8 of Integrated Investigations of Environmental Effects of Historical Mining in the Basin and Boulder Mining Districts, Boulder River Watershed, Jefferson County, Montana, D.A. Nimick, S.E. Church, and S.E. Finger, eds. U.S. Geological Survey Professional Paper 1652. Washington, DC: U.S. Government Printing Office [online]. Available: https://pubs.usgs.gov/pp/2004/1652/pdf/ChapD8.pdf [accessed June 23, 2017].

Church, S.E., P.B. von Guerard, and S.E. Finger. 2007. Integrated Investigations of Environmental Effects of Historical Mining in the Animas River Watershed, San Juan County, Colorado. Professional Paper 1651. U.S. Geological Survey [online]. Available: https://pubs.usgs.gov/pp/1651/ [accessed June 23, 2017].

Citeau, L. I. Lamy, F. van Oort, and F. Elsass. 2003. Colloidal facilitated transfer of metals in soils under different land use. Colloid. Surf. A 217(1-3):11-19.

Clark, H.F., D.J. Brabander, and R.M. Erdil. 2006. Sources, sinks, and exposure pathways of lead in urban garden soil. J. Environ. Qual. 35(6):2066-2074.

Clark, I., and P. Fritz. 1997. Environmental Isotopes in Hydrogeology. Boca Raton: Lewis.

Cloquet, C., J. Carignan, and G. Libourel. 2006. Isotopic composition of Zn and Pb atmospheric depositions in an urban/periurban area of northeastern France. Environ. Sci. Technol. 40(21): 6594-6600.

Cloy, J.M., J.G. Farmer, M.C. Graham, A.B. MacKenzie, and G.T. Cook. 2008. Historical records of atmospheric Pb deposition in four Scottish ombrotrophic peat bogs:

An isotopic comparison with other records from Western Europe and Greenland. Global Biogeochem. Cycles 2(2):GB2016.

Collins, A., D. Walling, and G. Leeks. 1997. Source type ascription for fluvial suspended sediment based on a quantitative composite fingerprinting technique. Catena 29(1):1-27.

Collins, A.L., D.E. Walling, H.M. Sichingabula, and G.J. Leeks. 2001. Suspended sediment source fingerprinting in a small tropical catchment and some management implications. Appl. Geogr. 21(4):387-412.

Collins, A., D. Walling, L. Webb, and P. King. 2009. Particulate organic carbon sources and delivery to river channels in the Somerset levels ECSFDI priority catchment, southwest UK. Int. J. River Basin Manag. 7(3):277-291.

Collins, A., D. Walling, L. Webb, and P. King. 2010. Apportioning catchment scale sediment sources using a modified composite fingerprinting technique incorporating property weightings and prior information. Geoderma 155(3-4):249-261.

Collins, A., Y. Zhang, D. McChesney, D. Waling, S. Haley, and P. Smith. 2012. Sediment source tracing in a lowland agricultural catchment in southern England using a modified procedure combining statistical analysis and numerical modelling. Sci. Total Environ. 414:301-317.

Collins, A., Y. Zhang, D. Duethmann, D. Walling, and K. Black. 2013. Using a novel tracing-tracking framework to source fine-grained sediment loss to watercourses at sub-catchment scale. Hydrol. Process. 27(6):959-974.

Cook, P.G., and J.K. Bohlke. 2000. Determining timescales for groundwater flow and solute transport. Pp. 1-30 in Environmental Tracers in Subsurface Hydrology, P.G. Cook, and A. Herczeg, eds. Boston: Kluwer.

Cooper, R.J., T. Krueger, K.M. Hiscock, B.G. Rawlins, 2014. Sensitivity of fluvial sediment source apportionment to mixing model assumptions: A Bayesian model comparison. Water Resour. Res. 50(11):9031–9047.

Croft, D.J., and K. Pye. 2004. Colour theory and the evaluation of an instrumental method of measurement using geological samples for forensic applications. Pp. 49-62 in Forensic Geosciences: Principles, Techniques, and Applications, K. Pye, and D.J. Croft, eds. Special Publication 232. London: The Geological Society of London.

Csavina, J., M.P. Taylor, O. Félix, K.P. Rine, A.E. Sáez, and E.A. Betterton. 2014. Size-resolved dust and aerosol contaminants associated with copper and lead smelting emissions: Implications for emissions management and human health. Sci. Total. Environ. 493:750-756.

Damman, A.W.H. 1986. Hydrology, development, and biogeochemistry of ombrogenous peat bogs with special reference to nutrient relocation in a western Newfoundland bog. Can. J. Bot. 64(2):384-394.

D'Amore, J.J., S.R. Al-Abed, K.G. Scheckel, and J.A. Ryan. 2005. Methods for speciation of metals in soils: A review. J. Environ. Qual. 34(5):1707-1745.

de Boer, D.H., and G. Crosby. 1995. Evaluating the potential of SEM/EDS analysis for fingerprinting suspended sediment derived from 2 contrasting topsoils. Catena 24(4):243-258.

D'Haen, K., G. Verstraeten, and P. Degryse. 2012. Fingerprinting historic fluvial sediment fluxes. Prog. Phys. Geogr. 36(2):154-186.

Douglas, G., M. Palmer, and G. Caitcheon. 2003. The provenance of sediments in Moreton Bay, Australia: A synthesis of major, trace element and Sr-Nd-Pb isotopic geochemistry, modeling and landscape analysis. Hydrobiologia 494(1):145-152.

Drivas, P., T. Bowers, and R. Yamartino. 2011. Soil mixing depth after atmospheric deposition. I. Model development and validation. Atmos. Environ. 45(25):4133-4140.

Du Laing, G., J. Rinklebe, B. Vandencasteele, E. Meers, and F.M.G. Tack. 2009. Trace metal behavior in estuarine and riverine floodplain sediments: A review. Sci. Total Environ. 407(13):3972-3985.

Eberl, D.D. 2004. Quantitative mineralogy of the Yukon River system: Changes with reach and season, and determining sediment provenance. Am. Miner. 89(11-12):1784-1794.

Eckel, W.P., M.B. Rabinowitz, and G.D. Foster. 2002. Investigation of unrecognized former secondary lead smelting sites: Confirmation by historical sources and elemental ratios in soil. Environ. Pollut. 117(2):273-279.

El Azzi, D., J. Viers, M. Guiresse, A. Probst, D. Aubert, J. Caparros, F. Charles, K. Guizien, and J.L. Probst. 2013. Origin and fate of copper in a small Mediterranean vineyard catchment: New insights from combined chemical extraction and δ65Cu isotopic composition. Sci. Total Environ. 463-464:91-101.

EPA (U.S. Environmental Protection Agency) 1996a. Method 1669: Sampling Ambient Water for Trace Metals at EPA Water Quality Criteria Levels. U.S. Environmental Protection Agency, Office of Water, Washington, DC [online]. Available: https://www.epa.gov/sites/production/files/2015-10/documents/method_1669_1996.pdf [accessed August 18th, 2017].

EPA. 1996b. EPA Method 3050B: Acid Digestion of Sediments, Sludges, and Soils; Revision 2. U.S. Environmental Protection Agency, Washington D.C. [online]. Available: https://www.epa.gov/homeland-security-research/epa-method-3050b-acid-digestion-sediments-sludges-and-soils [accessed August 18th, 2017].

EPA. 1998. Locating and Estimating Air Emissions from Sources of Lead and Lead Compounds. EPA-454/R-98-006. Office of Air Quality Planning and Standards, U.S. Environmental Protection Agency, Research Triangle Park, NC [online]. Available: https://www3.epa.gov/ttnchie1/le/lead.pdf [accessed July 3, 2017].

EPA. 1999a. MINTEQA2/PRODEFA2, A Geochemical Assessment Model for Environmental Systems: User Manual Supplement for Version 4.0 [online]. Available:

https://www.epa.gov/sites/production/files/documents/SUPPLE1.PDF [accessed July 3, 2017].

EPA. 1999b. Compendium of Methods for the Determination of Inorganic Compounds in Ambient Air. Compendium Method IO-3.5: Determination of Metals in Ambient Particulate Matter Using Inductively Coupled Plasma/Mass Spectrometry (ICP/MS). EPA/625/R-96/010a. US Environmental Protection Agency, Cincinnati, OH. June 1999 [online]. Available: https://www.epa.gov/sites/production/files/2015-07/documents/epa-io-3.5.pdf [accessed June 21, 2017].

EPA. 2007. Short Sheet: Estimating the Soil Lead Concentration Term for the Integrated Exposure Uptake Biokinetic (IEUBK) Model. OSWER 9200.1-78. U.S. Environmental Protection Agency, Office of Solid Waste and Emergency Response, U.S. Environmental Protection Agency, Washington, DC. September 2007 [online]. Available: https://semspub.epa.gov/work/HQ/175344.pdf [accessed June 21, 2017].

EPA. 2016. Receptor Modeling [online]. Available: https://www3.epa.gov/ttn/scram/receptorindex.htm [accessed July 3, 2017].

EPA. 2017a. Air Emission Modeling. SPECIATE Version 4.5 Through 4.0 [online]. Available: https://www.epa.gov/air-emissions-modeling/speciate-version-45-through-40 [accessed July 3, 2017].

EPA. 2017b. Air Emissions Inventories [online]. Available: https://www.epa.gov/air-emissions-inventories [accessed July 3, 2017].

Ettler, V. 2016. Soil contamination near non-ferrous metal smelters: A review. Appl. Geochem. 64:56-74.

Ettler, V., A. Vaněk, M. Mihaljevič, and P. Bezdička. 2005. Contrasting lead speciation in forest and tilled soils heavily polluted by lead metallurgy. Chemosphere 58(10):1449-1459.

Ettler, V., Z. Johan, B. Kříbek, F. Veselovský, M. Mihaljevič, A Vaněk, V. Penížek, V. Majer, O. Sracek, B. Mapani, F. Kamona, and I. Nyambe. 2016. Composition and fate of mine- and smelter-derived particles in soils of humid subtropical and hot semi-arid areas. Sci. Total Environ. 563-564:329-339.

Etyemezian, V.R., G. Nikolich, S.M. Ahonen, M.L. Pitchford, R. Sweeney, J. Purcell, J.A. Gillies, and H.D. Kuhns. 2007. The Portable In Situ Wind Erosion Laboratory (PI-SWERL): A new method to measure PM10 windblown dust properties and potential for emissions. Atmos. Environ. 41(18):3789-3796.

Evans, C., and T.D. Davies. 1998. Causes of concentration/discharge hysteresis and its potential as a tool for analysis of episode hydrochemistry. Water Resour. Res. 34(1):129-137.

Evrard, O., J. Poulenard, J. Nemery, S. Ayrault, N. Gratiot, C. Duvert, C. Prat, I. Lefevre, P. Bonte, and M. Esteves. 2013. Tracing sediment sources in a tropical highland catchment of central Mexico by using conventional and alternative fingerprinting methods. Hydrol. Process. 27(6):911-922.

Félix, O.I., J. Csavina, J. Field, K.P. Rine, E. Sáez, and E.A. Betterton. 2015. Use of lead isotopes to identify sources of metal and metalloid contaminants in atmospheric aerosol from mining operations. Chemosphere 122:219-226.

Ferrand, E., F. Eyrolle, O. Radakovitch, M. Provansal, S. Dufour, C. Vella, G. Raccasi, and R. Gurriaran. 2012. Historical levels of heavy metals and artificial radionuclides reconstructed from overbank sediment records in lower Rhône River (South-East France). Geochim. Cosmochim. Acta 82:163-182.

Foster, I.D.L., J.A. Less, A.R. Jones, A.S. Chapman, S.E. Turner, and R. Hodgkinson. 2002. The possible role of agricultural land drains in sediment delivery to a small reservoir, Worcestershire UK: A multiparameter tracing study. Pp. 443-442 in Structure, Function and Management Implications of Fluvial Sedimentary Systems, F.J. Dyer, M.C. Thoms, and J.M. Olley, eds. Wallingford, UK: IAHS Publications [online]. Available: https://www.researchgate.net/profile/Ian_Foster2/publication/268355050_The_possible_role_of_agricultural_land_drains_in_sediment_delivery_to_a_small_reservoir_Worcestershire_UK_A_multiparameter_fingerprint_study/links/5682ebec08ae19758391bd9d.pdf [accessed June 23, 2017].

Franks, S.W., and J.S. Rowan. 2000. Multi-parameter fingerprinting of sediment sources: Uncertainty estimation and tracer selection. Pp. 1067-1074 in Computational Methods in Water Resources, L.R. Bentley, C.A. Brebbia, W.G. Gray, G.F. Pinder, and J.F. Sykes, eds. Rotterdam: Balkema.

Foucher, D., N. Ogrinc, and H. Hintelmann. 2009. Tracing mercury contamination from the Idrija mining region (Slovenia) to the Gulf of Trieste using Hg isotope ratio measurement. Environ. Sci. Technol. 43(1):33-39.

Gallego, J.R., E. Rodreguez-Valdes, N. Esquinas, A. Fernandez-Brana, and E. Afif. 2016. Insights into a 20-ha multi-contaminated brownfield megasite: An environmental forensics approach. Sci. Total Environ. 563-564:683-692.

Gingele, F.X., and P. De Deckker. 2005. Clay mineral, geochemical and Sr-Nd isotopic fingerprinting of sediments in the Murray-Darling fluvial system, southeast Australia. Aust. J. Earth Sci. 52(6):965-974.

Gooseff, M.N., S.M. Wondzell, and K.E. Bencala. 2008. Solute transport along stream and river networks. Pp. 395-418 in River Confluences, Tributaries and the Fluvial Network, S. Rice, A. Roy, and B. Rhodes, eds. New York: John Wiley & Sons.

Graham, M.C., S.L. Vinogradoff, A.J. Chipchase, S.M. Dunn, J.R. Bacon, and J.G. Farmer. 2006. Using size fractionation and Pb isotopes to study Pb transport in the waters of an organic-rich upland catchment. Environ. Sci. Technol. 40(4):1250-1256.

Graney, J.R., and M.S. Landis. 2013. Coupling meteorology, metal concentrations and Pb isotopes for source attribution in archived precipitation samples. Sci. Total Environ. 448:141-150.

Graney, J.R., A.N. Halliday, G.J. Keeler, J.O. Nriagu, J.A. Robbins, and S.A. Norton. 1995. Isotopic record of lead pollution in lake sediments from the northeastern United States. Geochem. Cosmochim. Acta 59(9):1715-1728.

Graney, J.R., M.S. Landis, and S. Krupa. 2012. Coupling lead isotopes and element concentrations in epiphytic lichens to track sources of air emissions in the Athabasca Oil Sands Region. Pp. 343-372 in Alberta Oil Sands: Energy, Industry and the Environment, K. Percy ed., Amsterdam: Elsevier.

Gulson, B.L., K.G. Tiller, K.J. Mizon, and R.H. Merry. 1981. Use of lead isotope in soils to identify the source of lead contamination near Adelaide, South Australia. Environ. Sci. Technol. 15(6):691-696.

Guzmán, G., V. Barrn, and J.A. Gómez. 2010. Evaluation of magnetic iron oxides as sediment tracers in water erosion experiments. Catena 82:126-133.

Guzmán, G., J.N. Quinton, M.A. Nearing, L. Mabit, and J.A. Gómez. 2013. Sediment tracers in water erosion studies: Current approaches and challenges. J. Soil Sediment. 13(4):816-833.

Harbaugh, A.W. 2005. MODFLOW-2005, the U.S. Geological Survey Modular Ground-Water Model: The Ground-Water Flow Process. U.S. Geological Survey Techniques and Methods 6-A16 [online]. Available: https://pubs.usgs.gov/tm/2005/tm6A16/PDF.htm [accessed July 3, 2017].

Hardy, F., L. Bariteau, S. Lorrain, I. Theriault, G. Gagnon, D. Messier, D. Messier, and J.F. Rougene. 2010. Geochemical tracing and spatial evolution of the sediment bed load of the Romaine River, Quebec, Canada. Catena 81(1):66-76.

Harrison, J., H. Heijnis, and G. Caprarelli. 2003. Historical pollution variability form abandoned mine sites, Greater Blue Mountains World Heritage Area, New South Wales, Australia. Environ. Geol. 43(6):680-687.

Hayes, S.M., S.A. White, T.L. Thompson, R.M. Maier, and J. Chorover. 2009. Changes in lead and zinc lability during weathering-induced acidification of desert mine tailings: Coupling chemical and micro-scale analyses. Appl. Geochem. 42(12):2234-2245.

Hayes, S.M., S.M. Webb, J.R. Bargar, P.A. O'Day, R.M. Maier, and J. Chorover. 2012. Geochemical weathering increases lead bioaccessibility in semi-arid mine tailings. Environ. Sci. Technol. 46(11):5834-5841.

Heim, S., and J. Schwarzbauer. 2013. Pollution history revealed by sedimentary records: A review. Environ. Chem. Lett. 11(3):255-270.

Hopper, J.R., H.B. Ross, W.T. Sturges, and L.A. Barrie. 1991. Regional source discrimination of atmospheric aerosols in Europe using the isotopic composition of lead. Tellus 43(1):45-60.

Horowitz, A. 1991. A Primer on Sediment-Trace Element Chemistry, 2nd Ed. U.S. Geological Survey [online]. Available: https://pubs.usgs.gov/of/1991/0076/report.pdf [accessed June 28, 2017].

House, W.A., and M.S. Warwick. 1998. Hysteresis of the solute concentration/discharge relationship in rivers during storms. Water Res. 32(8):2279-2290.

Hudson-Edwards, K.A., M. Macklin, and M. Taylor. 1997. Historic metal mining input to Tees River sediment. Sci. Total Environ. 194-195:437-445.

Hudson-Edwards, K.A., M.G. Macklin, J.R. Miller, and P.J. Lechler. 2001. Sources, distribution and storage of heavy metals in the Rio Pilcomayo, Bolivia. J. Geochem. Explor. 72:229-250.

Hunt, A., D.L. Johnson, J.M. Watt, and I. Thornton. 1992. Characterizing the sources of particulate lead in house dust by automated scanning electron microscopy. Environ. Sci. Technol. 26(8):1513-1523.

Integral Consulting Inc. 2014. Sources of Lead to Residential Soils of St. Francois County. Report prepared for The Doe Run Company, May 15, 2014.

Jeong, J., and S.D. McDowell. 2003. Characterization and transport of contaminated sediments in the southern central Lake Superior. J. Miner. Mater. Charact. Eng. 2(2):111-135.

Johnson, C.E., T.G. Siccama, C.T. Driscoll, G.E. Likens, and R.E. Moeller. 1995. Changes in lead biogeochemistry in response to decreasing atmospheric inputs. Ecol. Appl. 5(3):813-822.

Kähkönen, M.A., K.P. Suominen, P.K. Manninen, and M.S. Salkinoja-Salonen. 1998. 100 years of sediment accumulation history of organic halogens and heavy metals in recipient and nonrecipient lakes of pulping industry in Finland. Environ. Sci. Technol. 32(12):1741-1746.

Keinonen, M. 1992. The isotopic composition of lead in man and the environment in Finland 1966–1987—Isotope ratios of lead as indicators of pollutant source. Sci. Total Environ. 113(3):251-268.

Kimball, B.A. 1997. Use of Tracer Injections and Synoptic Sampling to Measure Metal Loading From Acid Mine Drainage. U.S. Geological Survey Fact Sheet FS-245-296 [online]. Available: https://pubs.usgs.gov/fs/1996/0245/report.pdf [accessed June 28, 2017].

Kimball, B.A., and R.L. Runkel. 2009. Spatially detailed quantification of metal loading for decision making: Metal mass loading to American Fork and Mary Ellen Gulch, Utah. Mine Water Environ. 28(4):274-290.

Kimball, B.A., R.L. Runkel, K. Walton-Day, and K.E. Bencala. 2002. Assessment of metal loads in watersheds affected by acid mine drainage by using tracer injection and synoptic sampling: Cement Creek, Colorado, USA. Appl. Geochem. 17(9):1183-1207.

Kimball, B.A., R.L. Runkel, T.E. Cleasby, and D.A Nimick. 2005. Quantification of metal loading by tracer injection and synoptic sampling, 1997–1998. Pp. 191-261in Integrated Investigations of Environmental Effects of Historical Mining in the Basin and Boulder Mining Districts, Boulder River Watershed, Jefferson County, Montana, D.A. Nimick, S.E. Church, and S.E. Finger, eds. Professional Paper 1652. U.S. Geological Survey [online]. Available: https://pubs.usgs.gov/pp/2004/1652/pdf/ChapD6.pdf [accessed June 28, 2017].

Kimball, B.A., D.K. Nordstrom, R.L. Runkel, K.R. Vincent, and P.L. Verplanck. 2006. Questa Baseline and Pre-mining Groundwater Quality Investigation. 23. Quantification of Mass Loading from Mined and Unmined Areas along the Red River, New Mexico. Scientific Investigations Report 2006-5004. Salt Lake City, UT:U.S. Geological Survey [online]. Available: https://pubs.usgs.gov/sir/2006/5004/PDF/SIR2006_5004.pdf [accessed June 28, 2017].

Kimball, B.A., K. Walton-Day, and R.L. Runkel. 2007. Quantification of metal loading by tracer injection and synoptic sampling, 1996–2000. Pp. 417-495 in Integrated Investigations of Environmental Effects of Historical Mining in the Animas River Watershed, San Juan County, Colorado, S.E. Church, P.B. von Guerard, and S.E. Finger, eds. Professional Paper 1651. U.S. Geological Survey [online]. Available: https://pubs.usgs.gov/pp/1651/downloads/Vol1_combinedChapters/vol1_chapE9.pdf [accessed June 28, 2017].

Kimball, B.A., R.L. Runkel, and K. Walton-Day. 2010. An approach to quantify sources, seasonal change, and biogeochemical processes affecting metal loading in streams: Facilitating decisions for remediation of mine drainage. Appl. Geochem. 25(5):728-740.

Komárek, M., V. Ettler, V. Chrastny, and M. Milhaljevic. 2008. Lead isotopes in environmental sciences: A review. Environ. Int. 34(4):562-572.

Kuhns, H., V. Etyemezian, D. Landwehr, C. MacDougall, M. Pitchford, and M. Green. 2001. Testing re-entrained aerosol kinetic emissions from roads (TRAKER): A new approach to infer silt loadings on roadways. Atmos. Environ. 35(16):2815- 2825.

Kylander, M.E., D.J. Weiss, and B. Kober. 2009. Two high-resolution terrestrial records of atmospheric Pb deposition from New Brunswick, Canada, and Loch Laxford, Scotland. Sci. Total Environ. 407(5):1644-1657.

Linton, R.W., D.F.S. Natusch, R.L. Solomon, and C.A. Evans. 1980. Physicochemical characterization of lead in urban dusts. A microanalytical approach to lead tracing. Environ. Sci. Technol. 14(2):159-164.

Lloyd, C.E.M., J.E. Freer, P.J. Johnes, and A.L. Collins. 2016. Using hysteresis analysis of high-resolution water quality monitoring data, includng uncertainty, to infer controls on nutrient and sediment transfer in catchments. Sci. Total Environ. 543(Pt. A):388-404.

Lurry, D.L., and C.M. Kolbe. 2000. Interagency Field Manual for the Collection of Water-Quality Data. U.S. Geological Survey, Open-File Report 00-213 [online]. Available: https://pubs.usgs.gov/of/2000/ofr00-213/index.html [accessed September 21, 2017].

Ma, L., J. Konter, E.M. Herndon, L. Jin, G. Steinhoefel, D. Sanchez, and S.L. Brantley. 2014. Quantifying an early signature of the industrial revolution from lead concentrations and isotopes in soils of Pennsylvania, US. Anthropocene 7:16-29.

MacKinnon, G., A.B. MacKenzie, G.T. Cook, I.D. Pulford, H.J. Duncan, and E.M. Scott. 2011. Spatial and temporal variations in Pb concentrations and isotopic composition in road dust, farmland soil and vegetation in proximity to roads since cessation of use of leaded petrol in the UK. Sci. Total Environ. 409 (23):5010-5019.

Macklin, M.G., and R.B. Dowsett. 1989. The chemical and physical speciation of trace metals in fine grained overbank flood sediments in the Tyne Basin, north-east England. Catena 16(2):135-151.

Macklin, M.G., and K. Klimek. 1992. Dispersal, storage and transformation of metal-contaminated alluvium in the upper Vistula basin, southwest Poland. Appl. Geogr. 12(1):7-30.

Macklin, M.G., K.A. Hudson-Edwards, H.E. Jamieson, P. Brewer, T.J. Coulthard, A.J. Howard, and V.H. Renenda. 1999. Physical stability and rehabilitation of sustainable aquatic and riparian ecosystems in the Rio Guadiamar, Spain, following the Aznalcóllar mine tailings dam failure. Pp. 271-278 in Mine, Water and Environment, International Mine Water Association Congress, Seville, Spain [online]. Available: http://www.mwen.info/docs/imwa_1999/IMWA1999_Macklin_271.pdf [accessed June 28, 2017].

Macklin, M.G., P.A. Brewer, K.A. Hudson-Edwards, G. Bird, T.J. Coulthard, I.A. Dennis, P.J. Lechler, J.R. Miller, and J.N. Turner. 2006. A geomorphological approach to the management of rivers contaminated by metal-mining. Geomorphology 79(3-4):423-447.

Maher, B.A., S.J. Watkins, G. Brunskill, J. Alexander, and C.R. Fielding. 2009. Sediment provenance in a tropical fluvial and marine context by magnetic ‘fingerprinting’ of transportable sand fractions. Sedimentology 56(3):841-861.

Marcantonio, F., G.C. Flowers, and N. Templin. 2000. Lead contamination in a wetland watershed: Isotopes as fingerprints of pollution. Environ. Geol. 39(9):1070-1076.

Martínez-Carreras, N., T. Udelhoven, A. Krein, F. Gallart, J. Iffly, J. Ziebel, L. Hoffmann, L. Pfister, and D. Walling. 2010. The use of sediment colour measured by diffuse reflectance spectrometry to determine sediment sources: Application to the Attert River catchment (Luxembourg). J. Hydrol. 382(1-4):49-63.

Massoudieh, A., A. Gellis, W. Banks, and M. Wieczorek. 2013. Suspended sediment source apportionment in Chesapeake Bay watershed using Bayesian chemical mass balance receptor modeling. Hydrol. Process. 27(24):3363-3374.

Mattielli, N., J. Rimetz, J. Petit, E. Perdrix, K. Deboudt, P. Flament, and D. Weis. 2006. Zn-Cu isotopic study and speciation of airborne metal particles within a 5-km zone of a lead/zinc smelter. Geochim. Cosmochim. Acta 70(18):A401-A401 [online].Available:https://goldschmidtabstracts.info/2006/1372.pdf [accessed June 28, 2017].

Mattielli, N., J.C. Petit, K. Deboudt, P. Flament, E. Perdrix, A. Taillez, J. Rimetz-Planchon, and D. Weis. 2009. Zn isotope study of atmospheric emissions and dry depositions within a 5 km radius of a Pb-Zn refinery. Atmos. Environ. 43(6):1265-1272.

Matys Grygar, T., T. Novakoa, M. Mihaljevic, L. Strnad, and L. Koptikova. 2011. Surprisingly small increase in the sedimentation rate in the floodplain of Marova river in the Straznice area, Czech Republic, in the last 100 years. Catena 86:461-471.

Matys Grygar, T., J. Sedlacek, O. Bábek, T. Novakoa, L. Strnad, and M. Mihaljevic. 2012. Regional contamination of Moravia (south-eastern Czech Republic): Temporal shift of Pb and Zn loading in fluvial sediments. Water Air Soil Pollut. 223(2):739-753.

Matys Grygar, T.M., T. Nováková, O. Bábek, J. Elznicová, and N. Vadinová. 2013. Robust assessment of moderate heavy metal contamination levels in floodplain sediments: A cast study on the Jizera River, Czech Republic. Sci. Total Environ. 452-453:233-245.

McDiffett, W.F., A.W. Beidler, T.F. Dominick, and K.D. McCrea. 1989. Nutrient concentration-stream discharge relationships during storm events in a first order stream. Hydrobiologia 179(2):97-102.

Mihaljevič, M., V. Ettler, O. Šebek, and V. Chrastný. 2006. Lead isotopic signatures of wine and vineyard soils - Tracers of lead origin. J. Geochem. Explor. 88(1):130-133.

Mil-Homens, M., J. Blum, J. Canário, M. Caetano, A.M. Costa, S.M. Lebreiro, M. Trancoso, T. Richter, H. de Stigter, M. Johnson, and V. Branco. 2013. Tracing anthropogenic Hg and Pb input using stable Hg and Pb isotope ratios in sediments of the central Portuguese Margin. Chem. Geol. 336(16):62-71.

Miller, E.K., and A.J. Friedland. 1994. Lead migration in forest soils: Response to changing atmospheric inputs. Environ. Sci. Technol. 28(4):662–669.

Miller, J.R., and G. Mackin. 2013. Concentrations, sources, and potential ecological impacts of selected trace metals on aquatic biota within the Little Tennessee River Basin, North Carolina. Air Water Soil Pollut. 244:1613-1637.

Miller, J.R., and S.M. Orbock Miller. 2007. Contaminated Rivers: A Geomorphological-Geochemical Approach to Site Assessment and Remediation. Berlin: Springer.

Miller, J.R., M. Lord, S. Yurkovich, G. Mackin, and L. Kolenbrander. 2005. Historical trends in sedimentation rates and sediment provenance, Fairfield Lake, western North Carolina. JAWRA 41(5):1053-1075.

Miller, J.R., P.J. Lechler, G. Mackin, D. Germanoski, and L.F. Villarroel. 2007. Evaluation of particle dispersal from mining and milling operations using lead isotopic fingerprinting techniques, Rio Pilcomayo Basin, Bolivia. Sci. Total Environ. 384(1-3):355-373.

Miller, J.R., M. Lord, L.F. Villarroel, D. Germanoski, and J. Chambers. 2012. Structural organization of process zones in upland watersheds of Central Nevada and its influence on basin connectivity, dynamics, and wet meadow complexes. Geomorphology 139-140:384-402.

Miller, J.R., L. Wilson, P.J. Lechler, and L. Villarroel. 2013a. Utilization of geochemical tracers including Pb and Sb isotopes to determine trace metal sources within the Rio Loa Basin, Chile. Proceedings of the EnivonMetal Isotopes EMI2013, August 18-23, 2013, Ascona, Switzerland.

Miller, J.R., G. Mackin, P. Lechler, M. Lord, and S. Lorentz. 2013b. Influence of basin connectivity on sediment source, transport, and storage within the Mkabela Basin, South Africa. Hydrol. Earth Syst. Sci. 17:761-781.

Miller, J.R., G. Mackin, and S.M. Orbock Miller. 2015. Application of Geochemical Tracers to Fluvial Sediment. SpringerBriefs in Earth Sciences. Springer.

MO DNR (Missouri Department of Natural Resources). 2016. Air Pollution Control Program: 2016 Monitoring Network Plan [online]. Available: https://dnr.mo.gov/env/apcp/docs/2016monitoringnetworkplan.pdf [accessed June 28, 2017].

Morton, A.C. 1985. Heavy minerals in provenance studies. Pp. 249-277 in Provenance of Arenites, G.G. Zuffa, eds. Dordrecht, The Netherlands: Springer.

Müller, J., H. Ruppert, Y. Muramatsu, and J. Schneider. 2000. Reservoir sediments—A witness of mining and industrial development (Malter Reservoir, eastern Erzgebirge, Germany). Environ. Geol. 39(12):1341-1351.

NEPC (National Environmental Protection Council). 1998. National Environment Protection (Ambient Air Quality) Measure – Revised Impact Statement. National Environment Protection Council Service Corporation, Adelaide, Australia [online]. Available: http://www.ephc.gov.au/sites/default/files/AAQ_ImpStat__AAQ_NEPM_Revised_Impact_Statement_Final_199806.pdf [accessed June 28, 2017].

Nezat, C.A., S.A. Hatch, and T. Uecker. 2017. Heavy metal content in urban residential and park soils: A case study in Spokane, Washington, U.S.A. Appl. Geochem. 78:186-193.

Nimick, D.A., and T.E. Cleasby. 2001. Quantification of Metal Loads by Tracer Injection and Synoptic Sampling in Daisy Creek and the Stillwater River, Park County, Montana. Water-Resources Investigation Report 2000-

4261. U.S. Geological Survey [online]. Available: https://pubs.usgs.gov/wri/wri004261/ [accessed June 28, 2017].

Norrström, A., and G. Knutsson. 2012. Stable lead isotopes as tracers of groundwater pollution in the water supply for a small village. Environ. Earth Sci. 67(4):1085-1095.

Norton, S.A., E.R. Perry, T.A. Haines, and A.C. Dieffenbacher-Krall. 2004. Paleolimnological assessment of Grove and Plow Shop Ponds, Ayer, Massachusetts, USA – a superfund site. J. Environ. Monit. 6(5):457-465.

Nosrati, K., G. Govers, B.X. Semmens, and E.J. Ward. 2014. A mixing model to incorporate uncertainty in sediment fingerprinting. Geoderma. 217–218:173-180.

Olson, K.W., and R.K. Skogerboe. 1975. Identification of soil lead compounds from automotive sources. Environ. Sci. Technol. 9(3):227-230.

Ostergren, J.D., G.E. Brown, G.A. Parks, and T.N. Tingle. 1999. Quantitative speciation of lead in selected mine tailings from Leadville, CO. Environ. Sci. Technol. 33(10):1627-1636.

Padoan, M., E. Garzanti, Y. Harlavan, and I.M. Villa. 2011. Tracing Nile sediment sources by Sr and Nd isotope signatures (Uganda, Ethiopia, Sudan). Geochim. Cosmochim. Acta 75(12):3627-3644.

Palazón, L., B. Latorre, L. Gaspar, W.H. Blake, H.G. Smith, and A. Navas. 2015. Comparing catchment sediment fingerprinting procedures using an auto-evaluation approach with virtual sample mixtures. Sci. Total Environ. 532:456-66.

Parkhurst, D.L., and C.A.J. Appelo. 1999. User's Guide to PHREEQC (Version 2). A Computer Program for Speciation, Batch-reaction, OneDimensional Transport, and Inverse Geochemical Calculations. US Geological Survey Water-Resources Investigations Report 99-4259. U.S. Geological Survey [online]. Available: https://pubs.usgs.gov/wri/1999/4259/report.pdf [accessed August 10, 2017].

Pavlowsky, R.T., M.R. Owen, and D.J. Martin. 2010. Big River Mining Sediment Assessment Project: Distribution, Geochemistry, and Storage of Mining Sediment in Channel and Floodplain Deposits of the Big River System in St. Francois, Washington, and Jefferson Counties, Missouri, Final Report. OEWRI EDR-1-002. Ozarks Environmental and Water Resources Institute, Missouri State University, Springfield, MO.

Pease, P., S. Leece, P. Gares, and C. Rigsby. 2007. Heavy metal concentrations in sediment deposits on the Tary River floodplain following Hurricane Floyd. Environ. Geol. 51:1003-1111.

Petit, D., J.P. Mennesier, and L. Lamberts. 1984. Stable Pb isotopes in pond sediments as tracer of past and present atmospheric Pb pollution in Belgium. Atmos. Environ. 18(6):1189-1193.

Pittam, N., I. Foster, and T. Mighall. 2009. An integrated lake-catchment approach for determining sediment source changes at Aqualate Mere, Central England. J. Paleolimnol. 42(2):215-232.

Prapaipong, P., C.W. Enssle, J.D. Morris, E.L. Shock, and R.E. Lindvall. 2008. Rapid transport of anthropogenic lead through soils in southeast Missouri. Appl. Geochem. 23(8):2156-2170.

Pye, K. 2004. Forensic examination of rocks, sediments, soils, and dust using scanning electron microscopy and X-ray chemical microanalysis. Pp. 103-122 in Forensic Geosciences: Principles, Techniques, and Applications, K. Pye, and D.J. Croft, eds. Special Publication 232. London: Geological Society.

Rabinowitz, M.B., and G.W. Wetherill. 1972. Identifying sources of lead contamination by stable isotope techniques. Environ. Sci. Technol. 6(8):705-709.

Rehkämper, M., R. Wombacker, T.J. Horner, and Z. Xue. 2011. Natural and anthropogenic Cd isotope variations. Pp. 124-154 in Handbook of Environmental Isotope Geochemistry, M. Baskaran, ed. Heidelberg: Springer.

Rosenbauer, R.J., A.C. Foxgrover, J.R. Hein, and P.W. Swarzenski. 2013. A Sr-Nd isotopic study of sandsized sediment provenance and transport for the San Francisco Bay coastal system. Mar. Geol. 345:143-153.

Rowan, J.S., S. Black, and S.W. Franks. 2012. Sediment fingerprinting as an environmental forensics tool explaining cyanobacteria blooms in lakes. Appl. Geogr. 32(2):832-843.

Russell, M., D. Walling, and R. Hodgkinson. 2001. Suspended sediment sources in two small lowland agricultural catchments in the UK. J. Hydrol. 252(1-4):1-24.

Schaider, L.A., D.B. Senn, E.R. Estes, D.J. Brabander, and J.P. Shine. 2014. Sources and fates of heavy metals in a mining-impacted stream: Temporal variability and the role of iron oxides. Sci. Total Environ. 490:456-466.

Scheckel, K.G., J.A. Ryan, D. Allen, and N.V. Lescano. 2005. Determining speciation of Pb in phosphate-amended soils: Method limitations. Sci. Total Environ. 350(1-3):261-272.

Schreck, P., M. Schubert, K. Freyer, H.C. Treutler, and H. Weiss. 2005. Multi-metal contaminated stream sediment in the Mansfeld mining district: Metal provenance and source detection. Geochem. Expl. Environ. Anal. 5:51-57.

Shepherd, T.J., S.R.N. Chenery, V. Pashley, R.A. Lord, L.E. Ander, N. Breward, S.F. Hobbs, M. Horstwood, B.A. Klinck, and F. Worral. 2009. Regional lead isotope study of a polluted river catchment: River Wear, Northern England, UK. Sci. Total Environ. 407(17):4882-4893.

Sherman, L.S., J.D. Blum, J.T. Dvonch, L.E. Gratz, and M.S. Landis. 2015. The use of Pb, Sr, and Hg isotopes in Great Lakes precipitation as a tool for pollution source attribution. Sci. Total Environ. 502:362-274.

Shiel, A.E., D. Weis, and K.J. Orians. 2010. Evaluation of zinc, cadmium and lead isotope fractionation during smelting and refining. Sci. Total Environ. 408(11):2357-2368.

Shikazono, N., K. Tatewaki, K.M. Mohiuddin, T. Nakano, and H.M. Zakir. 2012. Sources, spatial variation, and speciation of heavy metals in sediments of the Tamagawa River in central Japan. Environ. Geochem. Health 34(Suppl. 1):13-26.

Shirahata, H., R.W. Elias, and C. Patterson. 1980. Chronological variations in concentrations and isotopic compositions and anthropogenic atmospheric Pb in sediments of a remote subalpine pond. Geochim. Cosmochim. Acta 44(2):149-162.

Shotyk, W. 1988. Review of the inorganic geochemistry of peats and peatland waters. Earth Sci. Rev. 25(2):95-176.

Shotyk, W., and M. Krachler. 2010. The isotopic evolution of atmospheric Pb in central Ontario since AD 1800, and its impacts on the soils, waters, and sediments of a forested watershed, Kawagama Lake. Geochim. Cosmochim. Acta 74(7):1963-1981.

Shotyk, W., D. Weiss, P.G. Appleby, A.K. Cheburkin, R. Frei, M. Gloor, J.D. Kramers, S. Reese, and V.O. Knaap. 1998. History of atmospheric lead deposition since 12,370 14C yr BP from a peat bog, Jura Mountains, Switzerland. Science 281(5383):1635-1640.

Simonetti A., C. Gariepy, and J. Carignan. 2000. Pb and Sr isotopic evidence for sources of atmospheric heavy metals and their deposition budgets in northeastern North America. Geochim. Cosmocim. Acta 64(20):3439-3452.

Sivry, Y., J. Riotte, J.E. Sonke, S. Audry, J. Schäfer, J. Viers, G. Blanc, R. Freydier, and B. Dupré. 2008. Zn isotopes as tracers of anthropogenic pollution from Zn-ore smelters The Riou Mort-Lot River system. Chem. Geol. 255(3-4):295-304.

Small, M.J., A.B. Nunn, B.L. Forslund, and D.A. Daily. 1995. Source attribution of elevated residential soil lead near a battery recycling site. Environ. Sci. Technol. 29(4):883-895.

Small, I.F., J.S. Rowan, and S.W. Franks. 2002. Quantitative sediment fingerprinting using a Bayesian uncertainty estimation framework. Pp. 443-450 in The Structure, Functions and Management Implications of Fluvial Sedimentory Systems, F.J. Dyer, M.C. Thoms, and J.M. Olley, eds. IAHS Publication no. 276. Wallingford, UK: IAHS Press.

Small, I.F., J.S. Rowan, S.W. Franks, A. Wyatt, and R.W. Duck. 2004. Bayesian sediment fingerprinting provides a robust tool for environmental forensic geosciences applications. Pp. 207-213 in Forensic Geosciences: Principles, Techniques, and Applications, K. Pye, and D.J. Croft. eds. Special Publication 232. London: The Geological Society of London.

Sobanska, S., N. Ricq, A. Laboudigue, R. Guillermo, C. Bremard, J. Laureyns, J.C. Merlin, and J.P. Wignacourt. 1999. Microchemical investigations of dust emitted by a lead smelter. Environ. Sci. Technol. 33(9):1334-1339.

Sonke, J.E., J.A. Hoogewerft, S.R. van der Laan, and J. Vangronsveld. 2002. A chemical and mineralogical reconstruction of Zn-smelter emissions in the Kempen region (Belgium), based on organic pool sediment cores. Sci. Total Environ. 292(1-2):101-119.

Sonke, J.E., Y. Sivry, J. Viers, R. Freydier, L. Dejonghe, L. Andre, J.K. Aggarwal, F. Fontan, and B. Dupre. 2008. Historical variations in the isotopic composition of atmospheric zinc deposition from a zinc smelter. Chem. Geol. 252(3-4):145-157.

Sonke, J.E., J. Schäfer, J. Chmeleff, S. Audry, G. Blanc, and B. Dupré. 2010. Sedimentary mercury stable isotope records of atmospheric and riverine pollution from two major European heavy metal refineries. Chem. Geol. 279(3-4):90-100.

Spiro, B., D.J. Weiss, O.W. Purvis, I. Mikhailova, B.J. Williamson, B.J. Coles, and V. Udachin. 2004. Lead isotopes in lichen transplants around a Cu smelter in Russia determined by MC-ICP-MS reveal transient records of multiple sources. Environ. Sci. Technol. 38(24):6522-6528.

Stafilov, T., R. Sajn, Z. Pencevski, B. Boev, M.V. Frontasyeva, and L.P. Strelkova. 2010. Heavy metal contamination of topsoils around a lead and zinc smelter in the Republic of Macedonia. J. Hazard. Mater. 175(1-3):896-914.

Steinmann, M., and P. Stille. 1997. Rare earth element behavior and Pb, Sr, Nd isotope systematics in a heavy metal contaminated soil. Appl. Geochem. 12(5):607-623.

Steinnes, E., T.E. Sjøbakk, C. Donisa, and M.L. Brännvall. 2005. Quantification of pollutant lead in forest soils. Soil Sci. Soc. Am. J. 69(5):1399-1404.

Sterling, D.A., D.L. Johnson, A.M. Murgueytio, and R.G. Evans. 1998. Source contribution of lead in house dust from a lead mining waste superfund site. J. Expo. Anal. Environ. Epidemiol. 8(3):359-379.

Stokes, S., and D. Walling. 2003. Radiogenic and isotopic methods for the direct dating of fluvial sediments. Pp. 231-267 in Tools in Fluvial Geomorphology, G.M. Kondolf, and H. Piégay, eds. Chichester, UK: John Wiley & Sons.

Sweeney, M., V.R. Etyemezian, T. Macpherson, W.G. Nickling, J.A. Gillies, G. Nikolich, and E. McDonald. 2008. Comparison of PI-SWERL with dust emission measurements from a straight-line field wind tunnel. J. Geophys. Res. 113:F01012.

Taylor, M.P., A.K. Mackay, K.A. Hudson-Edwards, and E. Holz. 2010. Soil Cd, Cu, Pb and Zn contaminants around Mount Isa city, Queensland, Australia: Potential sources and risks to human health. Appl. Geochem. 25(6):841-855.

Telmer, K., G.F. Bonham-Carter, D.A. Kliza, and G.E.M. Hall. 2004. The atmospheric transport and deposition of smelter emissions: Evidence from the multi-element geochemistry of snow, Quebec, Canada. Geochim. Cosmochim. Acta 68(14):2961-2980.

Tessier, A., P.G.C. Campbell, and M. Bisson. 1979. Sequential extraction procedure for the speciation of particulate trace metals. Anal. Chem. 51(7):844-851.

Thapalia, A., D.M. Borrok, P. Van Metre, M. Musgrove, and E.R. Landa. 2010. Zn and Cu isotopes as tracers of anthropogenic contamination in a sediment core from an urban lake. Environ. Sci. Technol. 44(5):1544-1550.

Tobin, G.A., R. Brinkmann, and B.E. Montz. 2000. Flooding and the distribution of selected metals in floodplain sediments in St. Maries, Idaho. Environ. Geochem. Health 22(3):219-232.

Ure, A.M., P. Quevauviller, H. Muntau, and B. Griepink. 1993. Speciation of heavy metals in soils and sediments. An account of the improvement and harmonization of extraction techniques undertaken under the auspices of the BCR of the Commission of the European Communities. Int. J. Environ. Anal. Chem. 51(1-4):135-151.

USGS (U.S. Geological Survey). 2006. National Field Manual for the Collection of Water-Quality Data. Chapter A4. Collection of Water Samples. Book 9 Handbooks for Water-Resources Investigations [online]. Available: https://water.usgs.gov/owq/FieldManual/chapter4/pdf/Chap4_v2.pdf [accessed July 3, 2017].

Villarroel, L.F., J.R. Miller, P.J. Lechler, and D. Germanoski. 2006. Lead, zinc, and antimony contamination of the Rio Chilco-Rio Tupiza drainage system, southern Bolivia. Environ. Geol. 51(2):283-299.

Von Gunten, H.R., M. Sturm, and R.N. Moser. 1997. 200-year record of metals in lake sediments and natural background concentrations. Environ. Sci. Technol. 31(8):2193-2197.

Walling, D.E., and J.C. Woodward. 1995. Tracing sources of suspended sediment in river basins: A case study of the river Culm, Devon, UK. Mar. Freshwater Res. 46(1):327-336.

Walton-Day, K., J.L. Flynn, B.A. Kimball, and R.L. Runkel. 2005. Mass Loading of Selected Major and Trace Elements in Lake Fork Creek Near Leadville, Colorado, September–October 2001. Scientific Investigation Repopt 2005-5151. U.S. Geological Survey [online]. Available: https://pubs.usgs.gov/sir/2005/5151/pdf/Walton-Day508.book.pdf [accessed June 28, 2017].

Wang, X., J.C. Chow, S.D. Kohl, K.E. Percy, A.H. Legge, and J.G. Watson. 2015. Characterization of PM2.5 and PM10 fugitive dust source profiles in the Athabasca Oil Sands Region. J. Air Waste Manage. Assoc. 65(12):1421-1433.

Watson, J.G., and J.C. Chow. 2005. Receptor models. Pp. 455-501 in Air Quality Modeling -Theories, Methodologies, Computational Techniques, and Available Databases and Software, Vol. II – Advanced Topics, P. Zannetti, ed. Pittsburgh, PA: Air and Waste Management Association and the EnviroComp Institute.

Watson, J.G., and J.C. Chow. 2013. Source apportionment. In Encyclopedia of Environmetrics, A.H. El-Shaarwi, and W.W. Piegorsch, eds. Chichester, UK: John Wiley & Sons.

Watson, J.G., J.C. and Chow. 2014. Source Apportionment. Wiley StatsRef: Statistics Reference Online.

Watson, J.G., and J.C. Chow. 2015. Receptor models and measurements for identifying and quantifying air pollution sources. Pp. 677-706 in Introduction to Environmental Forensics, 3rd Ed., B.L. Murphy, and R.D. Morrison, eds. Amsterdam, The Netherlands: Elsevier.

Watson, J.G., J.C. Chow, and C.A. Frazier. 1999. X-ray fluorescence analysis of ambient air samples. Pp. 67-96 in Elemental Analysis of Airborne Particles, Vol. 1, S. Landsberger, and M. Creatchman, eds. Amsterdam: Gordon and Breach Science.

Watson, J.G., T. Zhu, J.C. Chow, J.P. Engelbrecht, E.M. Fujita, and W.E. Wilson. 2002. Receptor modeling application framework for particle source apportionment. Chemosphere 49(9):1093-1136.

Watson, J.G., J.C. Chow, X. Wang, S.D. Kohl, and L.N.R. Yatavelli. 2014a. Windblown Fugitive Dust Characterization in the Athabasca Oil Sands Region. Prepared by Desert Research Institute, Reno, NV, for Wood Buffalo Environmental Association, Ft. McMurray, AB, Canada.

Watson, J.G., J.C. Chow, X. Wang, and S.D. Kohl. 2014b. Chemical Source Profiles for Geological Dust Samples From the Athabasca Oil Sands Region. Prepared by Desert Research Institute, Reno, NV, for Wood Buffalo Environmental Association, Ft. McMurray, AB, Canada.

Watson, J.G., J.C. Chow, G. Engling, L.W.A. Chen, and X.L. Wang. 2016. Source apportionment: Principles and methods. Pp. 72-125 in Airborne Particulate Matter: Sources, Atmospheric Processes and Health, R.M. Harrison, ed. London, UK: Royal Society of Chemistry.

Weiss, D., W. Shotyk, P.G. Appleby, J.D. Kramers, and A.K. Cheburkin. 1999. Atmospheric lead deposition since the industrial revolution recorded by five Swiss peat profiles: Enrichment factors, fluxes, isotopic composition, and sources. Environ. Sci. Technol. 33(9):1340-1352.

Weiss, D., W. Shotyk, S.E. Page, J.O. Rieley, S. Reese, and A. Martinez-Cortizas. 2002. The geochemistry of major and selected trace elements in a forested peat bog, Kalimantan, SE-Asia, and its implications on past atmospheric dust deposition. Geochim. Cosmochim. Acta 66(13):2307-2323.

Weiss, D.J, N. Rausch, T.F.D. Mason, B.J. Coles, J.J. Wilkinson, L. Ukonmaanaho, T. Arnold, and T.M. Nieminen. 2007. Atmospheric deposition and isotope biogeochemistry of zinc in ombrotropic peat. Geochim. Cosmochim. Acta 71(14):3488-3517.

Weldeab, S., K.C. Emeis, C. Hemleben, and W. Siebel. 2002. Provenance of lithogenic surface sediments and pathways of riverine suspended matter in the eastern Mediterranean Sea: Evidence from 143Nd/144Nd and 87Sr/86Sr ratios. Chem. Geol. 186(1-2):139-149.

Weltje, G. 2012. Quantitative models of sediment generation and provenance: State of the art and future developments. Sediment. Geol. 280:4-20.

Whittig, L.D., and W.R. Allardice. 1986. X-ray diffraction techniques. Pp. 331-362 in Methods of Soil Analysis, Part I. Physical and Mineralogical Methods, A. Klute, ed. Madison, WI: American Society of Agronomy-Soil Science Society of America.

Wirt, L., K.J. Leib, R. Melick, and D.J. Bove. 2001. Metal Loading Assessment of a Small Mountainous Sub-basin Characterized by Acid Drainage- Prospect Gulch, Upper Animas River Watershed, Colorado. Open-File Rep. 01-0258. U.S. Geological Survey [online]. Available: https://pubs.usgs.gov/of/2001/ofr-01-0258/ofr-01-0258.pdf [accessed June 28, 2017].

Yu, L.Z., and F. Oldfield. 1993. Quantitative sediment source ascription using magnetic measurements in a reservoir-catchment system near Nijar, S.E. Spain. Earth Surf. Proc. Land. 18(5):441-454.

Zhang, Y., A. Collins, and A. Horowitz. 2012. A preliminary assessment of the spatial sources of contemporary suspended sediment in the Ohio River basin, United States, using water quality data from the NASQAN programme in a source tracing procedure. Hydrol. Process. 26(3):326-334.

Zhu, C., and G. Anderson. 2002. Environmental Applications of Geochemical Modeling. Cambridge: Cambridge University Press.

Zota, A.R., L.A. Schaider, A.S. Ettinger, R.O. Wright, J.P. Shine, and J.D. Spengler. 2011. Metal sources and exposures in the homes of young children living near a mining-impacted Superfund site. J. Expo. Sci. Environ. Epidemiol. 21(5):495-505.

5

Implications for Southeast Missouri and Research Needs

In this report, the committee has described potential sources of lead contamination, transport mechanisms for environmental dispersal of lead, and investigative strategies for determining the sources of lead in environmental media. Here, the committee addresses specifically the implications of that information for the Southeast Missouri Lead Mining District, provides some research recommendations for optimizing approaches, and considers other opportunities for using some of the analytic techniques or methods for identifying potential sources of environmental lead contamination.

IMPLICATIONS FOR SOUTHEAST MISSOURI

Given the extensive mining history of Southeast Missouri, determining the sources of lead at sites in this region is complex. The presence of lead-ore deposits at or near the surface means that there are natural sources of lead at sites. Mining activities have produced waste materials, and the materials have been dispersed over wide areas, have been deposited many miles from their sources, and have come into contact with residential areas. A further complication is that lead can come from consumer products, including lead paint, gasoline, agricultural chemicals, batteries, and plumbing materials. In Chapter 4, the committee described various analytic methods and approaches for identifying sources of lead at a site and presented investigative strategies for implementing the various methods and approaches for soil, sediments, water, and air. As noted, the steps undertaken can depend on site characteristics and available resources. In the pages that follow, the committee first presents its overall conclusions regarding implementation of its strategies for Southeast Missouri and then provides two case studies that use data from the area to support its conclusions.

Overall Conclusions

Isotopic analytic techniques hold great potential to distinguish commercially processed lead or non-native ores from natural or locally mined materials in Southeast Missouri.[1] For example, compilation of published $^{206}Pb/^{207}Pb$ ratios in North American ore deposits, gasoline, and paints is described in the next section. On the basis of those results, it is likely that lead-isotope ratios will be able to be used to apportion the sources of lead in soils, sediments, or house dust between those commercial sources and natural or mined materials in the case of Southeast Missouri. One strength of using isotopic analysis is that it can pinpoint a source with high accuracy if there is a unique signature, as is the case, for example, in the lead-paint data described in the next section.

The use of the analytic methods in the investigative approaches to distinguish natural sources of lead from mined material is much more challenging and might not be possible, particularly for soil, in the absence of a unique isotopic signature. The natural and mined material often contains lead that exhibits similar isotopic values, and this inhibits the use of lead isotopes to determine source. It might therefore be necessary to use other methods alone or in combination as shown in the committee's investigative strategies to make a judgment about potential sources. Accordingly, determining the grain size, structure, and composition of soil particles might link them to a source. For example, analyzing structure and chemical composition could link soil particles to a smelting operation inasmuch as smelter-derived soil particles often contain spherical grains composed of sulfides, oxides, or glassy particles (Ettler 2016). The spatial distribution of the particles is also likely to vary as a function of wind direction and distance from the smelter; sulfides are relatively dense, and their concentration in soil is therefore likely to be higher closer to the smelter. Furthermore, differences in particles from tailings and chat piles might be determined partly on the basis of grain size; the former consists of relatively small grains (less than 74 μm), and the latter

[1]The committee uses the term *natural materials* to refer to the geologic background that includes the mineral deposits that have not been mined.

much larger grains (often 4–8 mm). In addition, concentrations of barium and barite might distinguish lead from the Southeast Missouri Barite District and the St. Francois County Mining Area. With time, however, weathering of soil minerals derived from primary ore or smelter sources might alter or dissolve them and potentially disperse lead as a sorbed species or substituted in minor amounts in new precipitated phases. The point is that the selection of specific characteristics for attribution must be based on physical, compositional, mineralogic, or geochemical differences in the various source materials, and it might be necessary to use multiple methods to differentiate between all sources present. Even when multiple analyses are used, however, source identification might not be definitive.

It might be possible to estimate the relative contributions of lead from sources by using data analysis. *Data analysis* is a catchall term for mathematical manipulations designed to extract useful patterns and narratives from empirically derived collections of numbers, or data sets. The assessment activities undertaken early in the Superfund process described in Chapter 1 typically generate substantial collections of such data. The field collection and laboratory analysis of samples follow quality-assurance protocols; consequently, the consistent quality of the resulting data makes them well suited to systematic analyses.

A concrete example of data analysis to help to establish the extent to which lead-based paint contributes to environmental lead contamination in the Southeast Missouri Lead Mining District is described as the second case study. The mass-balance analysis revealed that much less total lead is likely to reside in exterior paint on older houses than in the top 1 inch of soil in many Southeast Missouri residential yards. Consequently, the contribution of exterior paint is expected to be a small fraction of the average lead content measured in typical Southeast Missouri residential yards. Such data analysis can be used to determine the dominant contributors to the mass balance of lead at a given site. However, the committee emphasizes that lead paint could still be the most important source for human exposure. The committee did not examine exposure pathways to determine the most important source for human exposure.

Lead-Isotope Techniques: A Southeast Missouri Case Study

To use lead-isotope ratios effectively for source attribution, one must first determine the lead isotopic composition of the various potential sources. The technique works well if there is an isotopic contrast between the sources that one wishes to distinguish but can yield equivocal results if there are overlapping isotopic compositions in the various potential sources. In most cases, some potential sources can be eliminated, and the investigation can focus on fewer possible sources. Thus, the lead-isotope method should work for source attribution of lead in soil near residences in mining or smelting areas if the isotopic composition of the ores and tailings is distinct from that of other possible sources.

For this case study, the committee selected lead paint and gasoline as possible sources because widespread environmental contamination has resulted from their use. The committee assessed whether they could be distinguished from natural ores or tailings from Southeast Missouri. It is well known that the lead isotopic composition differs between various ore deposits (Stanton and Russel 1959) and between lead ingots and slags and contaminated soils associated with different mines (Rabinowitz 2002). Different gasoline producers and paint manufacturers used lead mined from different ore bodies and in some cases changed their sources through time (Rabinowitz and Hall 2002). Therefore, it is expected that in some specific cases the lead isotopic composition of gasoline and paint might differ from that of local ore bodies but in other cases might be the same and therefore not useful for source attribution. The lead ores from Southeast Missouri have an extremely unusual lead isotopic composition (Rabinowitz and Wetherill 1972; Goldhaber et al. 1995). Thus, the isotopic composition of lead mined from Southeast Missouri (and adjacent areas in the tristate region) appears to be distinct from lead mined in nearly all other major mines (and products made with that lead) in North America (Doe 1970). However, the local geologic background lead in Southeast Missouri also contains unusual lead, so it might not be possible to differentiate between background and mine-derived lead in Southeast Missouri on the basis of isotopes alone.

The committee compiled $^{206}Pb/^{207}Pb$ ratios for North American lead ores, gasoline, and paints from around the world. The compilation is shown in Figure 5-1. All the data are from peer-reviewed published articles except those from the Herculaneum smelter and nearby houses, which come from an exposure-investigation report prepared by the Agency for Toxic Substances and Disease Registry (Orloff 2002). In that report, two houses that were within 1 mile of the smelter were studied. The upper panel in the figure shows that Southeast Missouri ores, especially those from the Old Lead Belt, have considerably higher $^{206}Pb/^{207}Pb$ ratios than nearly all other North American ores that are plotted; only ores from the Upper Mississippi Valley District have overlapping lead-isotope ratios. In addition, 11 of the 12 samples of paint from the two houses sampled near the Herculaneum smelter in Southeast Missouri had $^{206}Pb/^{207}Pb$ ratios lower than dust, ore, and slag sampled at the nearby smelter, and one had a higher ratio. Thirty-one samples of paint pigment produced by six major US lead-paint manufacturers had ratios lower than dust, ore, and slag sampled at the smelter. The single high-ratio paint sample was from the exterior of the back door of a house within 1 mile of the Herculaneum smelter. Either that paint contained lead that was produced from Southeast Missouri ores or the back door was contaminated by lead emitted from the smelter. Lead concentrations of the paint were not given in Orloff (2002), so the possibility that the back-door paint had low lead concentration and that the isotopic composition reflects the addition of lead from the nearby smelter cannot be eliminated.

The lead-isotope ratio measurements of materials from near the Southeast Missouri smelter suggest that the source of lead in most or possibly all lead paint was not Southeast Missouri lead. The middle panel in Figure 5-1 shows the $^{206}Pb/^{207}Pb$ ratios for six studies of lead in paint sampled from houses in other US locations, including Massachusetts, California, and New Jersey. All the samples have $^{206}Pb/^{207}Pb$ ratios lower than those of Southeast Missouri ores; this supports the suggestion that most lead paint manufactured in the United States used lead from ore deposits that did not include Southeast Missouri ores. Finally, the lower panel of Figure 5-1 shows the $^{206}Pb/^{207}Pb$ ratios of house paint from France and Brazil, road paint from New York City, bridge paint from Australia, and gasoline; these samples have $^{206}Pb/^{207}Pb$ ratios below those of Southeast Missouri ores.

Figure 5-1 shows that the Southeast Missouri ores, especially Old Lead Belt ores, have higher $^{206}Pb/^{207}Pb$ ratios than other ore deposits, gasoline, or paints compiled for North America. As noted, the only exceptions are the ores from the Upper Mississippi Valley District, which were probably not produced in large quantities in the 20th century compared with those from Southeast Missouri, and the single paint sample from the house near the Herculaneum smelter. Thus, in the case of the Southeast Missouri Lead Mining District, lead-isotope ratios will likely be able to attribute the sources of lead in soils between mined materials and paints and gasoline in most cases. As discussed below, additional analyses of lead concentrations and isotopic compositions of exterior paints in older houses in Southeast Missouri will be needed to verify the range of lead isotopic composition of paints in the area and the origin of the single back-door paint that had a high $^{206}Pb/^{207}Pb$ ratio.

The committee emphasizes that Figure 5-1 contains $^{206}Pb/^{207}Pb$ ratios from a variety of sources that were measured with a variety of mass spectrometry methods. Thus, the accuracy of the measurements varies considerably between about 0.01% and 0.5%. However, the highest estimated error is still smaller than the symbol sizes on the diagram; thus, it is clear that even without high-precision measurements, lead in most paints should be easily distinguishable from Southeast Missouri ores. High-precision lead-isotope analyses have uncertainties that are about one-fiftieth of the value reflected by the size of the symbols in Figure 5-1; thus, it is likely that at specific sites near other ore bodies, there will be measurable differences in lead-isotope ratios between paint and ore materials. In Figure 5-1, the committee showed only $^{206}Pb/^{207}Pb$ ratios because only they were available in some data sets. However, it is well known that if plots of $^{206}Pb/^{204}Pb$, $^{207}Pb/^{204}Pb$ and $^{208}Pb/^{204}Pb$ are included, there can be greater resolving power for determining isotopic source contributions.

The committee notes that lead-isotope ratio measurements can be classified into three categories according to their precision: low-precision analyses of digests without chromatographic separation using quadrupole inductively coupled plasma mass spectrometry (ICP–MS), medium-precision analyses of digests without chromatographic separation using magnetic-sector ICP–MS, and high-precision analyses of digests after chromatographic separation using either multicollector magnetic-sector ICP–MS or thermal ionization mass spectrometry. Analysis costs vary widely among laboratories but increase from low to medium to high precision. The estimated cost per analysis is \$50–500 and depends on the sample type and the analytic method. In the case of the Old Lead Belt in Southeast Missouri, low- and medium-precision analyses are likely to be adequate because of the large difference in $^{206}Pb/^{207}Pb$ between paint and ore. However, high-precision analyses might in some cases allow differentiation between ores from the various different mineral deposits in the mining district. Furthermore, measurement of ^{204}Pb with high precision might be helpful for source apportionment in some cases. In general, the committee suggests that preliminary studies with medium precision should be conducted first to determine where there would be a benefit from high-precision analyses.

The committee notes that the utility of the lead isotopic methods in other mining and smelting regions in the United States has not been evaluated adequately and will depend on the lead isotopic composition of individual ore deposits and the isotopic contrast between the ores and other exposure sources, including paint. Ores from the Coeur d'Alene Idaho mining district have an unusually low $^{206}Pb/^{207}Pb$ ratio of about 1.06, which is lower than any of the paint or gasoline $^{206}Pb/^{207}Pb$ ratios that the committee compiled from the literature (Figure 5-1). Thus, in addition to the Southeast Missouri ores, Coeur d'Alene ores have $^{206}Pb/^{207}Pb$ ratios that contrast with those of paint and gasoline and should allow source-attribution studies. Furthermore, Vázquez Bahéna et al. (2017) found that in the Taxco mining district of southern Mexico the lead isotopic composition could be used to differentiate between local ores, glazed pottery, and paints. They concluded that women of childbearing age who had high blood lead and lived in the area had acquired lead predominantly from a mixture of local ores and pottery glaze.

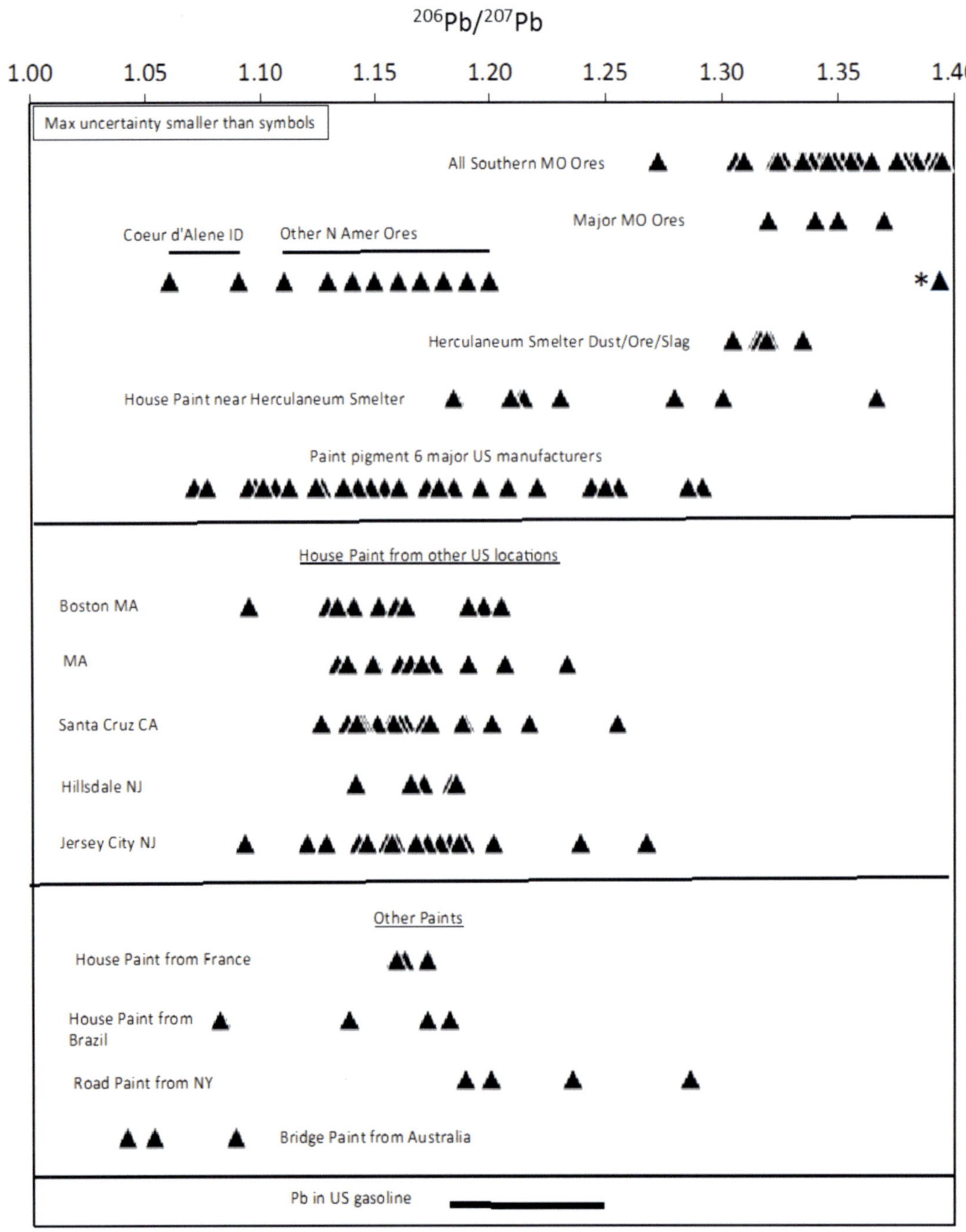

FIGURE 5-1 Compilation of $^{206}Pb/^{207}Pb$ ratios in North American ore deposits, gasoline, and all paints that have been published. Data are from the following sources: Southeast Missouri ores (Goldhaber et al. 1995); major Missouri ores and major North American mining districts (Rabinowitz 2005); dust, ore, and slag from the Herculaneum smelter and paint from a house near the Herculaneum smelter (Orloff 2002); paint pigments from six major US paint manufacturers (Rabinowitz and Hall 2002; in this study, only $^{206}Pb/^{204}Pb$ ratios were reported and were converted to $^{206}Pb/^{207}Pb$ ratios by using the linear relationship between $^{206}Pb/^{204}Pb$ and $^{206}Pb/^{207}Pb$ ratios for ingots, slags, and contaminated soils from a wide array of US lead ores and reported by Rabinowitz 2002); house paint from two houses within 1 mile of the Herculaneum smelter (Orloff 2001, 2002); Boston, MA, house paint (Rabinowitz (1995); Massachusetts (city not designated) house paint (Viczián et al.1990); Santa Cruz, CA, house paint (Gwiazda and Smith 2000); Hillsdale, NJ, house paint (Jaeger et al. 1998); Jersey City, NJ, house paint (Adgate et al. 1998); house paint from France (Glorennec et al. 2010); house paint from Brazil (Mirlean et al. 2005), road paint from New York City, NY (Duzgoren-Aydin and Weiss 2008), and bridge paint from Australia (Gulson et al. 2016); gasoline (Sherrell et al. 1992). The asterisked triangle represents the average $^{206}Pb/^{207}Pb$ ratios in ores from the Upper Mississippi Valley District (Heyl et al. 1974).

Data Analysis to Estimate Lead-Based Paint Contribution to Environmental Lead Contamination: A Southeast Missouri Case Study

As noted in Chapter 2, lead-based paints were heavily used in the early decades of the previous century (Mielke 1999), particularly to protect the exterior surfaces of houses and other structures from the effects of weathering. Detritus from lead-based paints is well recognized as a potential contaminant of soils in older neighborhoods, and protocols for the collection of residential soil samples distinguish drip zones where such debris is likely to collect adjacent to painted surfaces (see Figure 5-2). In addition to paint flakes and residues, drip zones can be enriched in airborne lead-containing particles deposited initially on roofs and walls and then washed off by precipitation. As Figure 5-2 indicates, samples collected in quadrants of the larger yard are intended to represent average concentrations in soils free from those influences.

Lead is considered to be relatively immobile as a constituent of soil in Southeast Missouri. Consequently, the spatial distribution of soil lead concentrations offers some indication of where exogenous lead might have been introduced and provides a strategy for attributing soil lead to point sources where releases are concentrated at fixed locations. The contributions of removal of lead-based paint in preparation for repainting and of weathering and wear of lead-based exterior paint are expected to manifest themselves as differentials in soil lead concentrations between the drip zone and the outer yard.

Figure 5-3 directly compares drip-zone with yard-average soil lead concentrations measured at 145 Southeast Missouri lots in the Focused Residential Sampling program described by Integral Consulting Inc. (2014). Drip-zone concentrations were found to be higher than yard concentrations at about 75% of homes (48 of 65) built before 1940, when lead content in house paint started to decline (Mielke 1999). The observation of higher drip-zone concentrations at older houses provides evidence that lead-based paint contributes to environmental contamination. Many other homes of all ages showed yard-average lead concentrations well above those in their drip zones, and these data indicate the presence of sources other than house paint. The mixture of positive and negative spatial concentration gradients near homes suggests that exterior paint does not generally overwhelm other sources of soil lead in the sampled neighborhoods.

A different kind of gradient that is evident in Figure 5-3 is the tendency of older houses to have higher soil lead concentrations outside their drip zones. Of the 60 residences built after 1979, 35 (58%) have yard lead concentrations below 400 mg/kg, whereas only six of 65 built

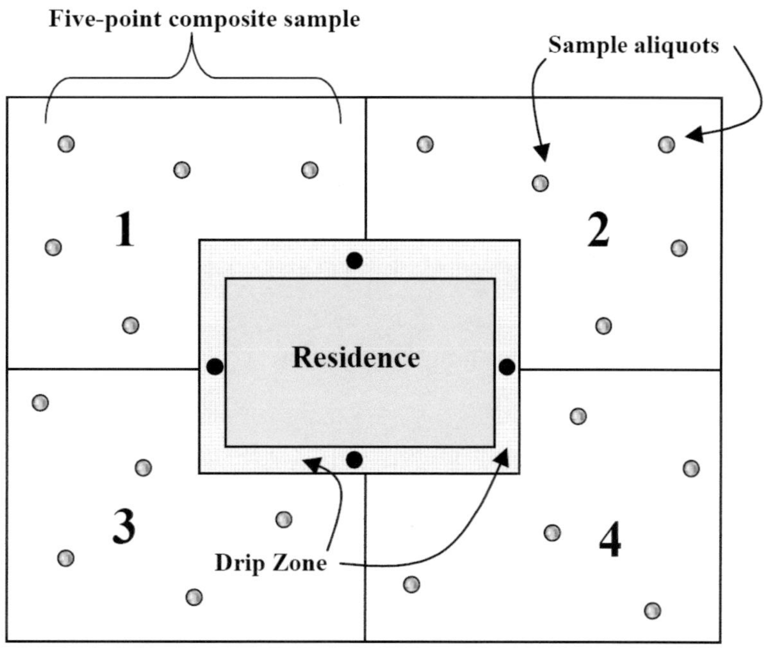

FIGURE 5-2 Recommended minimum soil sampling in yards greater than one-eighth acre. A five-point composite sample is collected 0–1 inch below the soil surface in each quadrant, outside the drip zone considered as extending 30 inches from the exterior walls of the house. Source: EPA (2003).

before 1940 (9%) are below this threshold. The differential could represent nothing more than the circumstance that construction at different times took place in different neighborhoods with different levels of prior contamination. Alternatively, it can be interpreted as evidence that lead-based paint debris is not confined to the drip zones of exterior walls. The higher lead concentrations recorded in older yards might be attributable in some part to the lead-based exterior paint used on the older houses, as suggested by Integral Consulting Inc. (2014, pp. 3-2, 4-1).

A mass balance can place bounds on source contributions of lead. Figure 5-3 compared *concentrations* in the outer yard with *concentrations* in the drip zone. The average lead concentration in the outer yard represents a much greater *total quantity* of lead, however, than would a similar concentration in the narrow drip zone. To assess the possibility raised above that lead-based paint is a major contributor to lead in the outer yard and not just to lead in the drip zone, the committee first considered the total lead contained in a yard's surface soil.

The total lead content of a yard's surface soil can be estimated from the average mass concentration and an estimate of soil volume and density. An arithmetic mean lead concentration of 677 mg/kg of soil was reported outside the drip zone for all Southeast Missouri yards sampled in 2013 (Integral Consulting Inc. 2014, Figure 3-8). The committee used the bulk soil density for the relevant density (Timm et al. 2005), which accounts for the volume contributed by interstitial air spaces. Surface soils vary in their composition and compaction; reported residential bulk soil densities typically are 1.2–1.6 g/cm^3. The committee took 1.5 g/cm^3 as a representative value, and this converts the mean lead concentration of 677 mg/kg reported for sampled Southeast Missouri yards to about 1 mg/cm^3 of soil. The committee assumed conservative values (lower bounds) of one quarter-acre for Southeast Missouri residential lot sizes, and 40% as an estimate of the portion covered with soil. A quarter-acre is about 10^7 cm^2, and the 40% soil-cover factor is offset by the sampling depth of 1 in (2.54 cm) rather than 1 cm. Those order-of-magnitude calculations yield average amounts of lead in the top inch of yard soil in a quarter-acre of at least 10 kg ([10^7 cm^3][1 mg/cm^3]). The 10-kg total lead burden estimated to reside in the top inch of an average Southeast Missouri yard serves as a meaningful benchmark with which the contents of other potential lead reservoirs can be compared. In the case of lead-based exterior paint, three methods described below can be used to estimate a house's total lead burden.

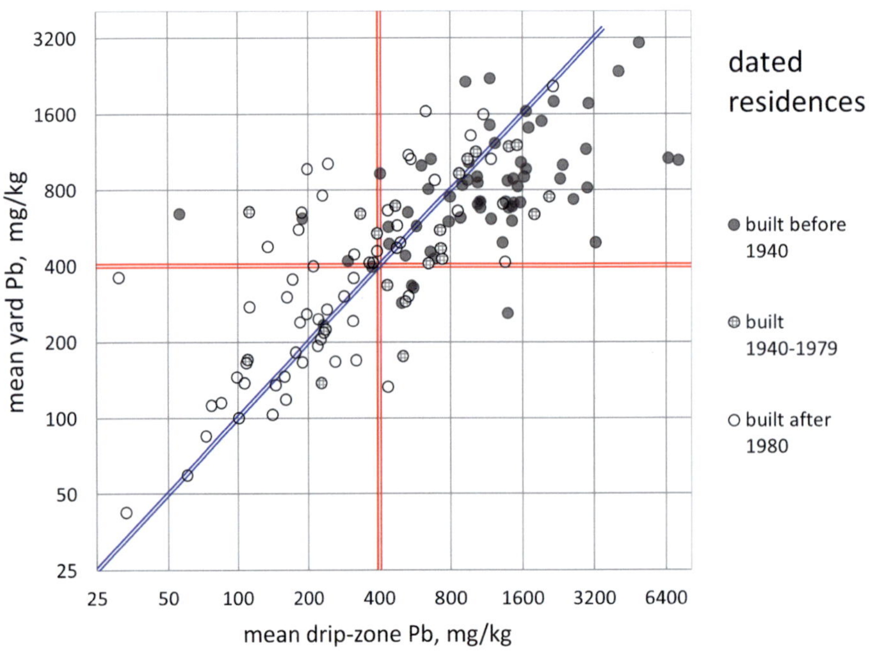

FIGURE 5-3 Results from 2013 soil sampling in Southeast Missouri. Each point represents one residential lot; its coordinates indicate drip-zone concentration (horizontal axis) and the arithmetic mean concentration of the four yard quadrants (vertical axis). The red lines indicate a typical action level for soil clean-up. Scales are logarithmic and marked at regular geometric intervals. The double blue line represents the yard concentrations equal to the drip-zone concentration. Soil lead concentrations are from Integral Consulting Inc. (2014). Only sampled residences (145 of 165) of which construction dates could be found on line from the real-estate data aggregator Zillow were plotted. Source: Generated from Zillow (2017).

• The lead content of house paints has varied, but a representative formulation might be the following (HUD 2012, Appendix A): 12% Pb × 50% solids × 12 lb/gal = lead at 0.7 lb/gal. A square house 40 ft on a side with walls that extend 9 ft above the foundation would have at most 1,440 ft^2 of paintable exterior surface. At coverages of 350–400 ft^2/gal, this house would take about 4 gal of paint, containing 2.8 lb = 1.3 kg. The 10-kg estimated in the surface yard soil would correspond to seven or eight full-coverage coats of lead-based paint.

• Mielke et al. (2001) studied a two-story wood frame house that was built in the 1920s and had deteriorating lead-based paint. The investigators oversaw an exterior restoration in which care was taken to collect and quantify all removed paint and reported 41 kg of paint scrapings that contained 3.7 kg of lead. They estimated that a similar amount remained bound intact to the wood, for a total of 7.4 kg of lead accumulated in paint layers over many decades.

• A third set of estimates comes from systematic national surveys undertaken for the Department of Housing and Urban Development and the US Environmental Protection Agency (EPA) by field inspectors using portable x-ray fluorescence analyzers (EPA 1995a). Table 5-1 summarizes the survey for exterior surfaces of occupied and privately owned housing units grouped into three construction periods. The historical decline of lead-based paint use is clear in the survey results. For the maximally leaded "pre-war" houses, the committee estimated total lead contents of

$$1{,}441 ft^2 \left(\frac{12\ in}{ft}\right)^2 \left(\frac{2.54\ cm}{in}\right)^2 4.6\frac{mg}{cm^2} = 6.2\ kg,$$

which is close to the value that Mielke et al. (2001) obtained by collecting paint scrapings.

The three estimates indicate that less total lead is likely to reside in exterior paint, even on an older house, than the 10-kg average lead burden that sampling indicates now sits in the top inch of sampled Southeast Missouri yards. Table 5-1 shows that the high-lead-content paints applied before 1940 were largely still in place when inspectors visited them more than 50 years later; that finding suggests that most of the paint ever applied to the exterior remains there. Some portion of the applied paint is lost to weathering, maintenance, and wear, but most of this debris is expected to accumulate in the drip zone, as discussed earlier.

The committee's mass-balance findings imply that the great majority of the surface lead in a randomly selected sample of surface yard soil outside the drip zone cannot reasonably be attributed to contamination by paint debris in typical Southeast Missouri yards. However, they do not preclude that paint debris might be primarily responsible for the soil lead content in some areas of some yards or that the paint debris is the primary source of human exposure. Again, the committee emphasizes that one must examine exposure pathways to identify the sources important for human exposure.

RESEARCH NEEDS

Additional research provides an opportunity to enhance the performance of the analytic methods incorporated into the investigative strategies proposed by the committee and to reduce uncertainty during implementation. The committee concluded that high research priority should be given to further characterizing potential sources but notes that the research described below is not essential for implementation of its investigative strategies. The research needs are not listed in order of priority.

Characterizing Lead Isotopic Composition of Historical Paints

The committee has discussed the use of lead-isotope ratios for attributing sources of lead in contaminated soils to mining and smelting activities. The committee pointed to examples of previous studies that showed an isotopic contrast between lead from paint and lead from mining and smelting activities in Southeast Missouri and suggested that the two sources of lead in soils could be identified by using lead isotopes. However, to evaluate the utility

TABLE 5-1 Selected Characteristics of Exterior Surfaces of Privately Owned Occupied Housing Units Built before 1980 with Lead-Based Paint

	Construction Era		
	Before 1940	1940–1959	1960–1979
Number of sampled units	77	87	120
Area of lead-based paint per unit (ft^2)	1,441	758	482
Mean lead-based paint loading (Pb in mg /cm^2)	4.6	1.5	0.6

Source: Extracted from EPA (1995b), Appendix II, Tables 2-1, 2-12, 2-13.

of the approach fully, additional study of lead isotopes should be conducted. In particular, knowledge of the lead isotopic composition of house paint in Southeast Missouri is limited to 11 analyses related to a single house and one analysis related to a second house, and lead concentrations in these samples were not reported. The committee recommends systematic sampling and lead isotopic analysis of paint in houses built before 1980 throughout the Southeast Missouri Lead Mining District. A broad sampling and analysis of paint in houses of various ages and locations should be conducted with at least five interior and five exterior paint samples collected from each house. That approach would give a much clearer indication of the range in lead isotopic composition of paint used in Southeast Missouri and would clarify the degree to which lead isotopes can be used in attributing lead in soil to paint vs mining and smelting sources. If there are other lead-containing consumer products at the site—such as plumbing materials, batteries, and pesticides—studies of their lead isotopic composition could establish or refute their role as a source of lead contamination at a site.

Characterizing Lead-Bearing Geologic Materials and Mining Wastes

The distribution, concentration, and geologic context of potentially key elements need to be better characterized to provide a baseline for interpretation of natural contributions to surface lead. There is a need to move away from just making bulk lead measurements and to provide greater detail on the materials and characteristics of the lead in the geologic and mining sources. Characterization would include chemical data, such as concentrations and isotopic ratios, mineralogy, grain size and texture, and host rock types. Spatial characterization would include vertical distribution, particularly in weathering (soil) profiles, and the lateral (areal) distribution over appropriate areas. Such information will improve confidence in investigative strategies for source attribution introduced in Chapter 4.

Recent advances in analytic chemistry, such as multicollector ICP–MS, have opened the possibility that a wide array of nontraditional stable isotopes can be used to determine the source of trace metals in the environment. Many metals—such as cadmium, copper, iron, mercury, and zinc—occur as sulfides and are associated with lead in geologic materials. Thus, they can potentially be used to attribute lead from geologic and mining sources in the Southeast Missouri Lead Mining District.

Variations in lead isotopic abundance in geologic materials are due mainly to radioactive decay over geologic timescales in that lead undergoes little change in isotopic ratios during other types of physical, chemical, and biologic processes. In contrast, shifts in the isotopic abundances of nontraditional stable isotopes[2] are small but are due entirely to physical, chemical, or biologic processes, including ones that occur during industrial processing or during metal dispersal at the Earth's surface (Wombacher et al. 2004; Cloquet et al. 2006; Shiel et al. 2010; Rehkämper et al. 2011; Blum et al 2014). It is that shift in the isotopic ratio that allows those isotopes to be of use in studies of source attribution. However, their effective use requires a detailed understanding of the processes that cause changes in the isotopic ratios in the environment of interest. They are more likely to change because of chemical reactions in the environment and are therefore more complicated to use for source attribution than lead isotopes. However, several such isotopes—including cadmium, mercury, and zinc—have proved useful as tracers of particles in soils and river sediments. Their application to the Southeast Missouri situation will require additional investigation but might prove to be highly useful, particularly for distinguishing between particles derived from ore deposits and smelter emissions.

Developing Capabilities to Model Metal Concentrations in Soils as a Function of Depth

As described in Chapter 4, the science of modeling sediment distributions in river floodplains is relatively advanced. Advantages of multielement, geochemical fingerprinting for attribution of sources of sediment and sediment-associated metals in rivers include (1) the ability to incorporate a wide array of elements into the analysis, thereby increasing the probability that the sources of sediment (and particulate lead) can be uniquely distinguished; (2) the establishment of widely accepted sampling and statistical analysis protocols; and (3) the increasing ability to quantify uncertainty in the attribution results (see, for example, Collins et al. 2012; Miller et al. 2015). The approach might prove applicable for determining particulate lead in sediments of the Big River with little modification of existing methods. Modeling of metal movement through soils is less developed. The application of multielemental, geochemical fingerprinting to determine sources of particulate lead in soils is more challenging because of potential alteration and dissolution of particulate lead. Theoretically, the approach is applicable to soils and might yield highly useful results, but the method will require additional testing when applied to soils. Studies could be completed on soils near known metal-emitting sources to map and model how the metals, especially lead, have moved through soils as a function of time.

[2]Nontraditional stable isotopes include elements other than carbon, hydrogen, oxygen, nitrogen, and sulfur. Examples are zinc, copper, molybdenum, mercury, iron, lithium, and cadmium.

Characterizing Size Distribution of Suspended Particles in Air Samples

Tailings impoundments, chat piles, and smelting emissions are important sources of airborne lead in a mining district; however, little is known about the size of suspended particles and their mobility. Knowledge of the size distribution of lead-containing airborne particulate matter is important for controlling particle transport and deposition. The largest particles have high terminal velocities and can settle within a few minutes to a few hours whereas particles whose diameters are a few micrometers can remain suspended for several days and can therefore be transported long distances. Size-selective sampling of air samples in the mining district should be conducted to determine the particle-size distribution and establish the $PM_{2.5}$ and PM_{10} levels. Size-selective sampling can be conducted upwind or downwind of a source, such as smelter or tailings pile, to evaluate contributions from the source. In addition to particle-size measurements, particle analysis by elemental, mineralogic, isotopic, and microscopic methods can be used to help to relate airborne and deposited particles to their sources.

Characterizing Rates of Emission of Airborne Particles from Mining Wastes

Little is known about rates of emission of airborne lead from stockpiles of mining wastes. Portable wind tunnels can be used to estimate the airborne particle emission rates and dust production from stockpiles of mining wastes, such as chat piles and tailings impoundments. Portable wind tunnels can also be useful in measuring the threshold wind velocity at a specific site that results in particle suspension and later transport downwind. This information can help to establish whether airborne lead from mining wastes is an important source of lead contamination downwind.

Identifying Sources with Investigative Tools in Industrial Archeology and Affiliated Disciplines

Industrial archeology is the systematic study of material evidence associated with historical industrial activities, such as resource extraction and raw-material processing; refining and manufacturing; storage, transport, and distribution; uses throughout product and byproduct life cycles; and waste disposal. Although historical industrial processes were generally on a smaller scale than modern industry, they operated less efficiently, were more likely to use processes and materials that produced hazardous byproducts, and lacked pollution controls. Thus, the relative pollution effect of historical industrial activities can be highly disproportionate relative to the scale of production. Understanding historical practices sheds light on their relative importance as a source of legacy pollution that persists long after the industries have disappeared.

Industrial archeology provides scientific methods for investigating the industrial life cycle of lead and drawing objective conclusions about how past practices throughout the industrial life cycle contributed to the present abundance, distribution, and forms of lead in the built and natural environments. Those investigative tools could help to identify sources of lead contamination, especially at large Superfund sites, including the Southeast Missouri Lead Mining District, and involving industrial archeologists and historians at the earliest stages of a remedial investigation could provide valuable insight on sources, transport, and fate of contaminants. The committee notes that the investigative tools of industrial archeology evolve as do the tools of any other discipline. A good resource is *IA*, the Journal of the Society for Industrial Archeology, which publishes on industrial archeology of industrial waste.

OTHER APPLICATIONS OF RECOMMENDED ANALYTIC TECHNIQUES: A CASE STUDY OF BLOOD LEAD AND ISOTOPIC ANALYSIS

The analytic techniques and methods incorporated into the committee's investigative strategies can be used for applications other than simply soil, sediments, water, and air. For example, isotopic analysis can be used for assessing sources of lead in biologic samples, such as blood and urine. When humans are exposed to lead, the lead isotopic composition in their blood and urine reflects the sources of lead exposure. Lead isotope investigations to determine the sources of lead in residents near the Herculaneum Lead Smelter in Southeast Missouri were reported by the Agency for Toxic Substances and Disease Registry and published as two exposure investigations (Orloff 2001, 2002). The committee summarized and interpreted the data in those reports as a case study to demonstrate both the power and the complexity of the use of lead isotopes for source attribution.

Two houses within 1 mile of the Herculaneum Lead Smelter in Southeast Missouri and their occupants were sampled, and the samples were analyzed for lead isotopic composition (Orloff 2001, 2002). The various environmental media and human specimens that were sampled and their lead isotopic compositions are shown in Figure 5-4. The committee focused on the data from house 1 because the sampling included 11 paint samples, compared with only one at house 2. The patterns in lead isotopic composition and the general conclusions are similar for the two houses.

As discussed above, lead ores and mining waste from the Southeast Missouri Lead Mining District have unusually high $^{206}Pb/^{207}Pb$ ratios compared with most other mining districts and therefore most commercial

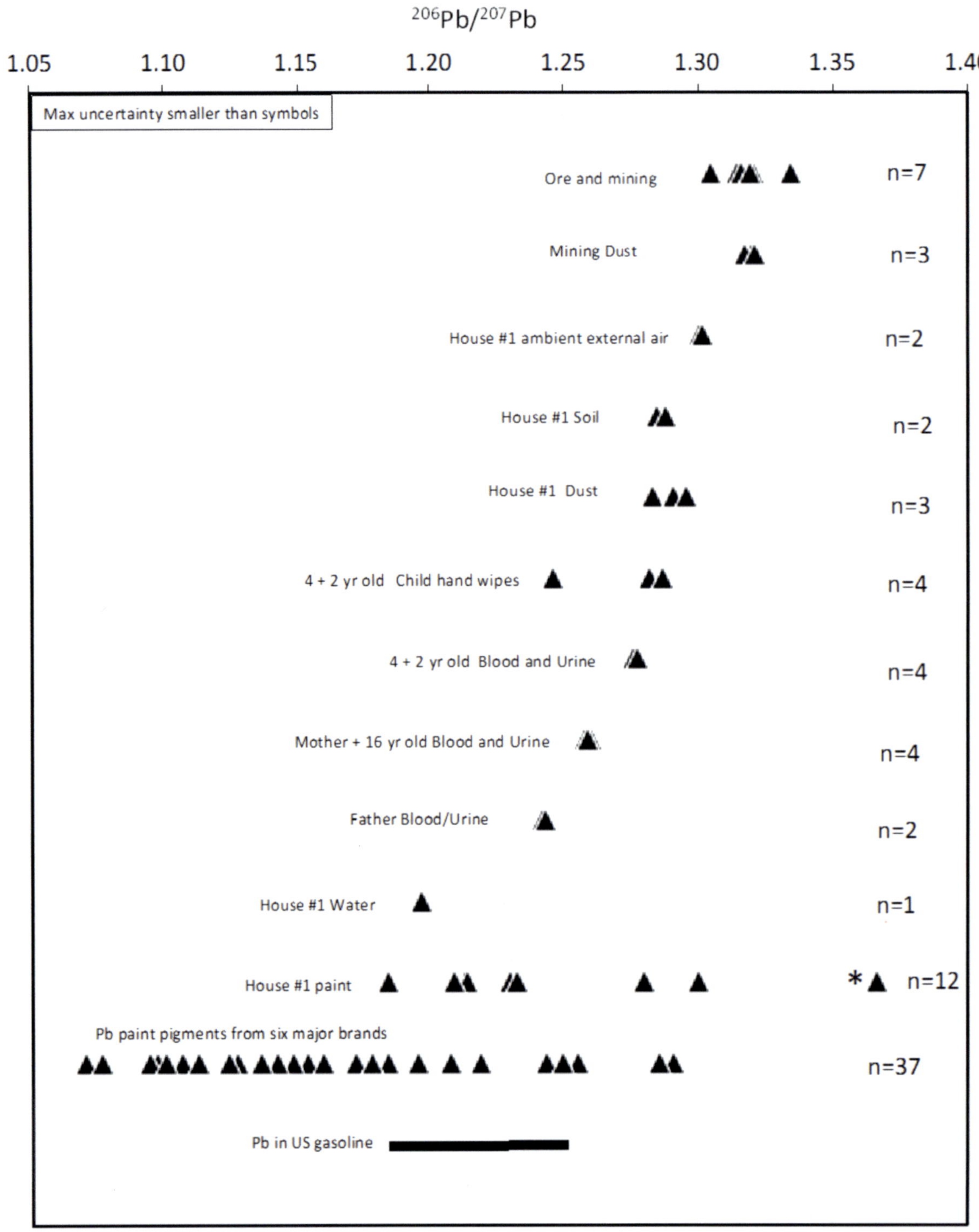

FIGURE 5-4 Compilation of $^{206}Pb/^{207}Pb$ ratios for one house within 1 mile of the Herculaneum Lead Smelter. House and its occupants were sampled, and the results were analyzed for lead isotopic composition (Orloff 2001, 2002). Also shown are $^{206}Pb/^{207}Pb$ ratios for ore and mined materials, mining dust, lead in paints and pigments of six major brands, and lead in gasoline.

products that contain lead. Ore, mining waste, and dust generated from ore-processing at the Herculaneum lead smelter have the highest $^{206}Pb/^{207}Pb$ ratios measured (Orloff 2001, 2002), 1.31–1.34. Ambient exterior-air particles larger than 0.8 μm sampled 4–5 ft above the ground in the house 1 yard have similar but slightly lower $^{206}Pb/^{207}Pb$ ratios, 1.30, and this suggests that the lead in atmospheric dust is generated mainly from the smelter but with the addition of some lower $^{206}Pb/^{207}Pb$ lead. Lead from leaded gasoline is a ubiquitous component of soils throughout North America. Although lead was phased out of automobile gasoline by the middle 1990s, it remains in soils, especially near roadways. As discussed in Chapter 2, leaded gasoline is still used in aviation fuel, which is currently the largest source of lead emitted to the atmosphere in the United States (Wolfe et al. 2016). The $^{206}Pb/^{207}Pb$ ratio of lead in gasoline in the United States has been estimated at 1.19–1.25 (Sherrell et al. 1992; Teutsch et al. 2001). The lead-isotope ratio of a mixture of sources whose lead-isotope ratios are known scales linearly with the proportion of lead from each source. As an example of how mixing calculations can be done, the committee used data from Figure 5-1 and applied a mixing calculation to explore the origin of the atmospheric dust samples analyzed by Orloff (2001). Using average $^{206}Pb/^{207}Pb$ ratios for each sample type, the committee found that the $^{206}Pb/^{207}Pb$ ratio can be explained by a mixture of about three-fourths smelter emissions and one-fourth lead from gasoline. However, the proportion of lead derived from gasoline could be much lower if the $^{206}Pb/^{207}Pb$ ratio of smelter emissions during the period of atmospheric sampling was at the lower end of the range of ratios of Southeast Missouri ores.

Soil from the yard and dust from the interior of house 1 have lower $^{206}Pb/^{207}Pb$ ratios than atmospheric dust, 1.28–1.30. The lower values could be produced by mixing smelter lead with lead from house paint. Eleven samples of interior and exterior house paint from house 1 were analyzed. Ten had values that overlap the previously reported range of ratios of commercial paint pigments (Rabinowitz and Hall 2002), 1.19–1.30. One paint sample has a $^{206}Pb/^{207}Pb$ ratio above the range of major lead ores; it is an exterior sample, and its lead concentration was not reported, so its lead isotopic composition is difficult to interpret.

As expected from other studies, blood and urine samples from each family member yielded nearly identical $^{206}Pb/^{207}Pb$ ratios. The 2- and 4-year-old children have $^{206}Pb/^{207}Pb$ ratios of 1.28—at the lower range of soil, dust, and hand-wipe values—except for the one hand-wipe outlier that might be related to paint exposure, perhaps from a windowsill. That result suggests that the 2-year-old and 4-year-old are exposed to lead mainly via ingestion of soil and dust; this result has been found to be the case in many other studies (see, for example, Gulson et al. 1994; Gwiazda and Smith 2000; Manton et al. 2000; Laidlaw et al. 2016). The 16-year-old and the mother have lower $^{206}Pb/^{207}Pb$ ratios, 1.26, and the father has a still lower ratio, 1.24. Those results are not unexpected in that even in an adolescent blood lead can reflect a mixture of past and present lead exposure. That is, lead is stored in bone material, and blood lead can be derived partly from lead mobilization from bone over year-long timescales (Manton 1985; Manton et al. 2000).

This case study demonstrates the power of lead isotopes in differentiating between lead sources in several types of environmental and biologic media. The identification of the lead sources for a particular sample is sometimes equivocal, but some sources can be eliminated in most cases, thus narrowing the possible combination of sources that need to be considered. In the Southeast Missouri Lead Mining District examined here, the largest source of lead isotopic uncertainty is the multiple values in paint at different locations in house 1. Many paint samples analyzed in Orloff (2002) were from locations where a child would not typically be exposed to them, including above a ceiling, on a baseboard molding, and in a parent's bedroom closet. By limiting paint sampling to areas where children have access and to where paint dust is generated by friction on door jambs or window sashes, the $^{206}Pb/^{207}Pb$ ratio of paint in a given house can be estimated more narrowly. Peeling or flaking paint that can be ingested by a child presents a different problem: the amount ingested can lead to frank lead poisoning or, in a worst-case scenario, encephalopathy that results in permanent brain damage or death. Even with the uncertainties and imperfect sampling strategy, Orloff (2001, 2002) demonstrates systematic differences of lead sources from different environmental sources and allows the testing and narrowing of hypotheses for the sources of human lead exposure.

Other studies that have used lead-isotope ratios to determine potential sources of lead in children in the United States (Viczián et al. 1990; Rabinowitz 1995; Gwiazda and Smith 2000; Manton et al. 2000; Glorennec et al. 2010) have been conducted in urban areas, not near mining or smelter sites. Researchers measured lead isotopes in children's blood and then sought to distinguish between potential sources, including house paints, household dust, and soils. Most of the studies involved some individual houses where children's blood-lead isotopes match with paint or soil and some other houses where there is such a large range in paint lead-isotope ratios that no single source can be identified. One such study was carried out in Santa Cruz, CA, by Gwiazda and Smith (2000). They measured the lead isotopic composition of children's blood, household dust, exterior soil, and interior and exterior paints. In one house, paint could be eliminated as a major source of lead exposure, and exterior dust and soil were implicated. In two other houses, paint from some areas could be eliminated as possible sources, but various mixtures of paints from other areas with dust and soil could explain the children's blood-lead isotope values, and no single source could be identified.

If EPA or others want to pursue blood-lead isotopic investigations for the Southeast Missouri Lead Mining District, the type of investigation described in this case study would need to be replicated in houses of different ages and at different locations in Southeast Missouri to determine where and under what conditions the lead isotope method can provide unambiguous attribution of lead sources for children who have high blood lead concentrations. The approach is promising in Southeast Missouri because the lead associated with mining and smelting has an unusually high $^{206}Pb/^{207}Pb$ ratio, about 1.34, which in most cases is distinct from lead used to make paint in the United States.

REFERENCES

Adgate, J.L., G.G. Rhoads, and P.J. Lioy. 1998. The use of isotope ratios to apportion sources of lead in Jersey City, NJ, house dust wipe samples. Sci. Total Environ. 221(2-3):171-180.

Blum, J.D., L.S. Sherman, and M.V. Johnson. 2014. Mercury isotopes in earth and environmental science. Annu. Rev. Earth Planet. Sci. 42:249-269.

Cloquet, C., J. Carignan, and G. Libourel. 2006. Isotopic composition of Zn and Pb atmospheric depositions in an urban/periurban area of northeastern France. Environ. Sci. Technol. 40(21):6594-6600.

Collins, A., Y. Zhang, D. McChesney, D. Waling, S. Haley, and P. Smith. 2012. Sediment source tracing in a lowland agricultural catchment in southern England using a modified procedure combining statistical analysis and numerical modelling. Sci. Total Environ. 414:301-317.

Doe, B.R. 1970. Lead Isotopes. Berlin: Springer.

Duzgoren-Aydin, N.S., and E.A.L. Weiss. 2008. Use and abuse of Pb-isotope fingerprinting technique and GIS mapping data to assess lead in environmental studies. Environ. Geochem. Health 30(6):577-588.

EPA (U.S. Environmental Protection Agency). 1995a. Report on the National Survey of Lead-Based Paint in Housing, Base Report. EPA 747-R95-003. U.S. Environmental Protection Agency, Office of Pollution Prevention and Toxics, Washington, DC [online]. Available: https://www.epa.gov/sites/production/files/documents/r95-003.pdf [accessed June 21, 2017].

EPA. 1995b. Report on the National Survey of Lead-Based Paint in Housing, Appendix II. Analysis. EPA 747-R95-005. U.S. Environmental Protection Agency, Office of Pollution Prevention and Toxics, Washington, DC [online]. Available: https://www.epa.gov/sites/production/files/documents/r95-005.pdf [accessed June 21, 2017].

EPA. 2003. Superfund Lead-Contaminated Residential Sites Handbook. OSWER 9285.7-50. U.S. Environmental Protection Agency, Lead Sites Workgroup [online]. Available: https://semspub.epa.gov/work/HQ/175343.pdf [accessed June 21, 2017].

Ettler, V. 2016. Soil contamination near non-ferrous metal smelters: A review. Appl. Geochem. 64:56-74.

Glorennec, P., C. Peyr, J. Poupon, Y. Oulhote, and B. Le Bot. 2010. Identifying sources of lead exposure for children, with lead concentrations and isotope ratios. J. Occup. Environ. Hyg. 7(5):253-260.

Goldhaber, M., S.E. Church, B.R. Doe, I.N. Aleinikoff, J.C. Brannon, F.A. Podosek, E.L. Mosier, C.D. Taylor, and C.A. Gent. 1995. Lead and sulfur-isotope investigations of Paleozoic sedimentary rocks from the southern mid-continent of the United States: Implications for the paleohydrology and ore genesis of the southeast Missouri lead-belt. Econ. Geol. 90(7):1875-1910.

Gulson, B., K.J. Mizon, A.J. Law, M.J. Korsch, J.J. Davis, and D. Howarth. 1994. Source and pathways of lead in humans from the Broken Hill Mining Community – An alternative use of exploration methods. Econ. Geol. 89(4):889-908.

Gulson, B., M. Chiaradia, J. Davis, and G. O'Connor. 2016. Impact on the environment from steel bridge paint deterioration using lead isotopic tracing, paint compositions and soil deconstruction. Sci. Total Environ. 550:69-72.

Gwiazda, R.H., and D.R. Smith. 2000. Lead isotopes as a supplementary tool in the routine evaluation of household lead hazards. Environ. Health Perspect. 108(11):1091-1097.

Heyl, A.V., G.P. Landis, and R.E. Zartman. 1974. Isotopic evidence for the origin of Mississippi Valley type mineral deposits: A review. Econ. Geol. 69(6):992-1006.

HUD (U.S. Department of Housing and Urban Development). 2012. Guidelines for the Evaluation and Control of Lead-Based Paint Hazards in Housing, 2nd Ed., Appendix 1. Units of Measure. U.S. Department of Housing and Urban Development, Office of Healthy Homes and Lead Hazard Control [online]. Available: https://portal.hud.gov/hudportal/HUD?src=/program_offices/healthy_homes/lbp/hudguidelines [accessed June 21, 2017].

Integral Consulting, Inc. 2014. Sources of Lead to Residential Soils of St. Francois County, Appendix A. Lead Copper and Zinc Concentrations in Residential Soils, 2013 Focused Residential Sampling. Report prepared for The Doe Run Company, May 15, 2014.

Jaeger, R.J., A.L. Weiss, and W.I. Manton. 1998. Isotopic ratio analysis in residential lead-based paint and associated surficial dust. Clin. Toxicol. 36(7):691-703.

Laidlaw, M.A.S., G.M. Filippelli, R.C. Sadler, C.R. Gonzales, A.S. Ball, and H.W. Mielke. 2016. Children's blood lead seasonality in Flint, Michigan (USA), and soil-sourced lead hazard risks. Int. J. Environ. Res. Public Health 13(4):358; doi:10.3390/ijerph13040358.

Manton, W.I. 1985. Total contribution of airborne lead to blood lead. Br. J. Ind. Med. 42(3):168-172.

Manton, W.I., C.R. Angle, K.L. Stanek, Y.R. Reese, and T.J. Kuehnemann. 2000. Acquisition and retention of lead by young children. Environ. Res. 82(1):60-80.

Mielke, H.W. 1999. Lead in the inner cities: Policies to reduce children's exposure to lead may be overlooking a major source of lead in the environment. Am. Sci. 87(1):62-73.

Mielke, H.W., E.T. Powell, A. Shah, C.R. Gonzales, and P.W. Mielke. 2001. Multiple metal contamination from house paints: Consequences of power sanding and paint scraping in New Orleans. Environ. Health Perspect. 109(9):973-978.

Miller, J.R., M. Mackin, and S.M. Orbock Miller. 2015. Application of Geochemical Tracers to Fluvial Sediment. New York: Springer. 142 pp.

Mirlean, N., D. Robinson, K. Kawashita, M.L. Vignol, R. Conceição, and F. Chemale. 2005. Identification of local sources of lead in atmospheric deposits in an urban area in Southern Brazil using stable lead isotope ratios. Atmos. Environ. 39(33):6204-6212.

Orloff, K.G. 2001. Exposure Investigation: Herculaneum Lead Smelter Site, Herculaneum, Jefferson County, Missouri, EPA Facility MOD006266373. September 14, 2001.

Orloff, K.G. 2002. Exposure Investigation: Herculaneum Lead Smelter Site, Herculaneum, Jefferson County, Missouri, EPA Facility MOD006266373. December 12, 2002 [online]. Available: http://health.mo.gov/living/environment/hazsubstancesites/pdf/Leadsourcesinchildren2002.pdf [accessed June 21, 2017].

Rabinowitz, M.B. 1995. Stable Isotopes of lead for source identification. Clin. Toxicol. 33(6):649-655.

Rabinowitz, M. 2002. Isotopic characterization of various brands of corroding grade refined lead metal. Bull. Environ. Contam. Toxicol. 69(4):501-508.

Rabinowitz, M.B 2005. Lead isotopes in soils near five historic American lead smelters and refineries. Sci. Total Environ. 346(1-3):138-148.

Rabinowitz, M.B., and G.S. Hall. 2002. Isotopic characterization of six major brands of white basic lead carbonate paint pigments. Bull. Environ. Contam. Toxicol. 69(5):617-623.

Rabinowitz, M.B., and G.W. Wetherill. 1972. Identifying sources of lead contamination by stable isotope techniques. Environ. Sci. Technol. 6(8):705-709.

Rehkämper, M., R. Wombacker, T.J. Horner, and Z. Xue. 2011. Natural and anthropogenic Cd isotope variations. Pp. 124-154 in Handbook of Environmental Isotope Geochemistry, M. Baskaran, ed. Heidelberg: Springer.

Sherrell, R.M., E.A. Boyle, and B. Hamelin. 1992. Isotopic equilibration between dissolved and suspended particulate lead in the Atlantic Ocean: Evidence from 210Pb and stable Pb isotopes. J. Geophys. Res. 97(C7):11,257-11,268.

Shiel, A.E., D. Weis, and K.J. Orians. 2010. Evaluation of zinc, cadmium and lead isotope fractionation during smelting and refining. Sci. Total Environ. 408(11):2357-2368.

Stanton, R.L., and R.D. Russell. 1959. Anomalous leads and the emplacement of lead sulfide ores. Econ. Geol. 54(4):588-607.

Teutsch, N., Y. Erel, L. Halicz, and A. Banin. 2001. Distribution of natural and anthropogenic lead in Mediterranean soils. Geochim. Cosmochim. Acta 65(17):2853-2864.

Timm, L.C., L.F. Pires, K. Reichardt, R. Roveratti, J. Oliveira, and P Bacchi. 2005. Soil bulk density evaluation by conventional and nuclear methods. Aust. J. Soil Res. 43:97-103.

Vázquez Bahéna, A.B., O. Talavera Mendoza, M.E. Moreno Godínez, S.A. Salgado Souto, J. Ruiz, and G. Huerta Beristain. 2017. Source apportionment of lead in the blood of women of reproductive age living near tailings in Taxco, Guerrero, Mexico: An isotopic study. Sci. Total Environ. 583:104-114.

Viczián, M., A. Lásztity, and R.M. Barnes. 1990. Identification of potential environmental sources of childhood lead poisoning by inductively coupled plasma mass spectrometry. Verification and case studies. J. Anal. At. Spectrom. 5(4):293-300.

Wolfe, P.J., A. Giang, A. Ashok, N.E. Selin, and S.R.H. Barrett. 2016. Costs of IQ loss from leaded aviation gasoline emissions. Environ. Sci. Technol. 50(17)9026-9033.

Wombacher, F., M. Rahkämper, and K. Mezger. 2004. Determination of the mass-dependence of cadmium isotope fractionation during evaporation. Geochim. Cosmochim. Acta 68(1):2349–2357.

Zillow. 2017. How much is My Home Worth? [online]. Available: https://www.zillow.com/how-much-is-my-home-worth/ [accessed June 21, 2017].

A

Challenges to Obtaining Information about Lead Sources at Superfund Sites Associated with Lead-Mining Districts

Source identification is not typically problematic for sites that have concentrated waste materials, but challenges arise when sites have large volumes of low-concentration wastes, such as large municipal landfills and mining sites. Such sites are sometimes referred to as *special sites*, and the US Environmental Protection Agency (EPA) remedial investigation and feasibility study (RI/FS) guidance "does not specifically state how... such sites should be addressed" (EPA 1988a, Section 1.5).

Lack of clear RI/FS guidance creates later challenges for assigning response costs at special sites among responsible parties and challenges to obtaining information about lead-source identification. Response costs at Superfund sites are divided among responsible parties by apportionment or allocation. Apportionment applies only when the liability of responsible parties can be separated; it would apply, for example, at a site where two parties are responsible for separate types of pollution that can be clearly distinguished qualitatively and quantitatively. Allocation applies at most Superfund sites because it is the default process and because separate liability is difficult to establish. At sites where allocation is used, equitable factors are used to assign liability for response costs. Although the Comprehensive Environmental Response, Compensation, and Liability Act (CERCLA, Superfund) does not prescribe the equitable factors to be used, Gore factors[1] are often applied and include the following:

- Ability of the parties to demonstrate that their contributions to a discharge, release, or disposal of a hazardous waste can be distinguished.
- Amount of hazardous waste involved.
- Degree of toxicity of hazardous waste involved.
- Degree of involvement by the parties in generation, transportation, treatment, storage, or disposal.
- Degree of care exercised by the parties with respect to the hazardous wastes concerned.
- Degree of cooperation with public officials to prevent harm to public health or to the environment.

Other equitable factors that have been used include contractual provisions, years of and economic benefit from ownership or operation, knowledge of and responsibility for waste handling and disposal practices, ability to pay, and strength of evidence that identifies responsible parties (Allen 2006).

EPA's orphan-share policies and guidance (EPA 1996a,b, 1997) and mixed-funding guidance (EPA 1987, 1988b) create additional challenges to identifying the sources of lead associated with Superfund sites. Orphan-share policies, such as the work policy (EPA 1996a) and the cost-recovery policy (EPA 1997), attempt to address the liability for a Superfund site that is attributed to insolvent or defunct parties that cannot pay their share of a site's cleanup cost. Here, EPA agrees to pay a portion of the orphan share allocated to insolvent or defunct parties; this can accelerate settlement with the viable responsible parties (EPA 2001). Mixed-funding is another mechanism whereby the federal government can accept settlement offers for less than the full response cost at a site. Mixed funding is appropriate under two circumstances: when pursuit of a fully funded settlement might cost the federal government more than accepting a mixed-fund settlement and when a substantial portion of the waste at a site cannot be attributed to known and financially viable parties. By creating opportunities to shift responsibility for response costs, the orphan-share policies and mixed-funding guidance create disincentives to objective disclosure of information about sources of lead at Superfund sites that are associated with lead-mining districts.

Liability for response costs at special sites, including Superfund sites associated with lead-mining districts, can range from hundreds of millions to billions of dollars. A party's ability or inability to demonstrate its contributions to a discharge, release, or disposal of a hazardous waste can substantially alter the outcome of the allocation process, orphan-share settlement, and mixed-fund settlement. The magnitude of the costs that are subject to allocation creates powerful incentives for potentially responsible parties and the federal government to think and

[1]H.R. 7020, 96th Cong., 2d Sess. (1980).

act strategically when it comes to gathering and sharing information about the sources of contamination at Superfund sites, particularly special sites. That situation creates impediments to open development and sharing of source-identification methods and results.

REFERENCES

Allen, D. 2006. Cost Allocations – Apportioning Environmental Response Costs Among Multiple Parties for the Cleanup of Contaminated Property. Montaq, March 23, 2006 [online]. Available: http://www.mondaq.com/unitedstates/x/38572/landlord+tenant+leases/Cost+Allocations+Apportioning+Environmental+Response+Costs+Among+Multiple+Parties+for+the+Cleanup+of+Contaminated+Property [accessed June 29, 2017].

EPA (U.S. Environmental Protection Agency). 1987. Evaluating Mixed Funding Settlements Under Cercla. October 20, 1987 [online]. Available: https://www.epa.gov/sites/production/files/2013-10/documents/mixfnd-cercla-mem.pdf [accessed June 28, 2017].

EPA. 1988a. Guidance for Conducting Remedial Investigations and Feasibility Studies. EPA540/G–89/004. Office of Emergency and Remedial Response, U.S. Environmental Protection Agency, Washington, DC. October 1988 [online]. Available: https://semspub.epa.gov/work/HQ/128301.pdf [accessed June 28, 2017].

EPA. 1988b. Interim Policy on Mixed Funding Settlements Involving the Preauthorization of States or Political Subdivision [online]. Available: https://www.epa.gov/sites/production/files/2013-10/documents/mixfund-preau-mem.pdf [accessed June 29, 2017].

EPA. 1996a. Interim Guidance on Orphan Share Compensation for Settlors of Remedial Action and Non-Time-Critical Removals [online]. Available: https://www.epa.gov/sites/production/files/2013-10/documents/orphan-share-rpt.pdf [accessed June 28, 2017].

EPA. 1996b. Orphan Share Reform: EPA Increasing Fairness in the Enforcement Process. Fact Sheet, June 4, 1996 [online]. Available: https://www.epa.gov/sites/production/files/2013-10/documents/orphan-shrref-rpt.pdf [accessed July 11, 2017].

EPA. 1997. Addendum to the "Interim CERCLA Settlement Policy Issued on December 5, 1984" [online]. Available: https://www.epa.gov/sites/production/files/2013-10/documents/adden-settle-mem.pdf [accessed June 29, 2017].

EPA. 2001. Orphan Share Superfund Reform Questions and Answers [online]. Available: https://www.epa.gov/sites/production/files/2013-10/documents/orph-sh-ref-qa.pdf [accessed June 29, 2017].

B

Biographic Information on the Committee on Sources of Lead Contamination at or near Superfund Sites

Edward J. Bouwer (*Chair*) is Abel Wolman Professor of Environmental Engineering in the newly formed Department of Environmental Health and Engineering at Johns Hopkins University. He is also director of the Center for Contaminant Transport, Fate and Remediation. Dr. Bouwer has extensive experience with drinking water and wastewater treatment processes, microbial process engineering, and contaminant transport and fate. His research interests encompass factors that influence biotransformation of contaminants; bioremediation for control of contaminated soils and groundwater; biofilm kinetics; biologic process design in wastewater, industrial, and drinking-water treatment; transport and fate of micro-organisms in porous media; behavior of metals in contaminated sediments; and defining and managing environmental risks. He serves as the editor-in-chief for *Biodegradation* and on the editorial boards for the *Journal of Contaminant Hydrology and Environmental Engineering Science*. Dr. Bouwer received his BS in civil engineering with a minor in nuclear engineering from Arizona State University, and an MS and a PhD in environmental engineering and science from Stanford University.

Mark D. Barton is a professor of geosciences at the University of Arizona. He serves as associate director of the Lowell Institute for Mineral Resources. He has been instrumental in driving innovative research applied to ore deposits in North America and South America for nearly 30 years. Collaborative research in which he has a leading role includes porphyry copper life cycles in the American Southwest and its broader crustal context, the regional geologic framework and origin of Carlin-type gold deposits, the characteristics and origin of iron oxide mineralization, granite petrogenesis and lithophile element mineralization in the Great Basin, and metallogenic synthesis in Mexico and adjoining regions. He is a winner of the Society of Economic Geology Lindgren Award and of the Mineralogical Society of America award and the author of numerous papers, including landmark reviews on ore deposits and metallogeny. He served on the National Research Council Committee on Basic Research Opportunities in the Earth Sciences. Dr. Barton earned his PhD in geology from the University of Chicago.

Eric A. Betterton is a University Distinguished Professor and head of the Department of Hydrology and Atmospheric Sciences at the University of Arizona. He also has appointments in the Department of Chemical and Environmental Engineering and the College of Public Health. His research focuses on air pollution, especially atmospheric aerosols and dust associated with mining operations; solar energy; rain, snow, and ice chemistry; remediation of polluted groundwater; and environmental fate of various compounds. He received a PhD in chemical kinetics from The University of Witwatersrand, South Africa.

Joel D. Blum holds the JD MacArthur Professorship and is the GJ Keeler Distinguished University Professor in the Department of Earth and Environmental Sciences of the University of Michigan. He obtained his BA from Case Western Reserve University, his MS from the University of Alaska, and his PhD from the California Institute of Technology. He was a professor of earth sciences at Dartmouth College for 9 years before moving to the University of Michigan in 1999. Dr. Blum's research focuses on the sources, fate and cycling of metals in the environment. Most recently, he has focused on understanding the biogeochemistry of mercury in aquatic, terrestrial, and atmospheric systems, in many cases using mercury stable isotopes. Dr. Blum has published over 200 research papers. He has served on numerous advisory panels for the National Science Foundation, the National Institutes of Health, the Environmental Law and Policy Center, and the Department of Energy. He is a past editor of *Chemical Geology* and current editor of *ACS Earth and Space Chemistry*. He is a fellow of the Geochemical Society, the

American Geophysical Union, the Geological Society of America, and the American Association for the Advancement of Science. In 2013, he won the Patterson Medal of the Geochemical Society for his work on metals in the environment.

Susan L. Brantley is Distinguished Professor of Geosciences and director of the Earth and Environmental Systems Institute at Pennsylvania State University. Dr. Brantley investigates chemical and physical processes associated with the circulation of aqueous fluids in shallow hydrogeologic settings. Her investigations incorporate field and laboratory work and theoretical modeling of observations. Particular interests are the measurement and prediction of the rates of natural processes, including chemical weathering and natural degassing. Her recent work has focused on measuring and modeling how rock turns into regolith and on water-quality issues in areas with hydraulic fracturing. Dr. Brantley has served on numerous National Research Council committees and is a member of the National Academy of Sciences. She is a recipient of the Wollaston Medal of the Geological Society, the Arthur L. Day Medal of the Geological Society of America, and the Presidential Award of the Soil Science Society of America. She received her PhD in geologic and geophysical sciences from Princeton University.

Judith C. Chow holds the Nazir and Mary Ansari Chair in Science and Entrepreneurialism and is a research professor in the Division of Atmospheric Sciences of the Desert Research Institute of the Nevada System of Higher Education in Reno, Nevada. For more than 39 years, she has conducted air-quality studies and performed data analysis to improve understanding of effects of air quality on human health, visibility, historical treasures, ecosystems, and climate. She has been principal investigator or a major collaborator in more than 50 large air-quality studies and many smaller ones in the United States and other countries. Dr. Chow is a charter member of EPA's Clean Air Scientific Advisory Committee (CASAC) and a member of CASAC's Particulate Matter Review Panel (2015–2018). She is the principal author or co-author of more than 400 peer-reviewed journal articles and book. She was chair of the Publications Committee for the *Journal of the Air & Waste Management Association* and serves on editorial boards or as thematic editor for several international journals, including the *Air Quality, Atmosphere, & Health, Aerosol* and *Air Quality Research, Atmospheric Pollution Research,* and *Particuology*. Dr. Chow has received the California Air Resources Board's 2011 Haagen-Smit Clean Air Award for her contributions to air-quality science and technology, the Air & Waste Management Association's 2016 Arthur C. Stern Distinguished Paper Award and 2002 Frank A. Chambers Excellence in Air Pollution Control Award, and the 2001 Nevada System of Higher Education's Nevada Regents' Researcher Award. Dr. Chow earned her ScD in environmental science and physiology from Harvard University.

Scott E. Fendorf is the Huffington Family Professor in Earth Sciences at Stanford University and a senior fellow at the Stanford Woods Institute for the Environment. Dr. Fendorf is interested in the chemical and biologic processes that govern the fate and transport of contaminants and nutrients in soils, sediments, and surface waters. His research group examines the chemical environments that develop as a result of biotic and abiotic processes and strives to account for the physical complexity in natural settings. Dr. Fendorf served on the National Research Council Committee on Bioavailability of Contaminants in Soils and Sediments. He received his PhD in soil and environmental chemistry from the University of Delaware.

Robert E. Hazen is a research scientist at the New Jersey Department of Environmental Protection in the Division of Science Research and Environmental Health. His research activities have focused on human health risk assessment and natural-resource management with the goal of assisting with department procedures and regulations. He has served on numerous national technical committees, including the US Environmental Protection Agency Science Advisory Board and the National Research Council Committee on Remedial Action Priorities for Hazardous Waste Sites. He received his PhD in biology and environmental health from New York University.

Chris E. Johnson is a professor in the Department of Civil and Environmental Engineering at Syracuse University. Dr. Johnson's research focuses on the chemistry of natural organic matter, which plays an important role in soil fertility, trace-metal transport, and the acid–base status of soils and natural waters. He is also involved in the interdisciplinary study of biogeochemical cycles at the Hubbard Brook Experimental Forest in the United States. He has research interests in the fate of trace metals (lead, zinc, copper, and nickel) in forest soils and landscapes; the effects of clear-cut logging on soils and drainage waters; and the changing acid–base chemistry of soils historically affected by acid rain. He served on the US Environmental Protection Agency's Clean Air Scientific Advisory Committee Lead Review Panel in 2011-2013, which established the National Ambient Air Quality Standard for lead. He has been an associate editor for the *Soil Science Society of America Journal* and the *Journal of Soils and Sediments*. Dr. Johnson has a BSE in civil and urban engineering, an MA in statistics, and a PhD in geology from the University of Pennsylvania.

William I. Manton is a professor of geology at the University of Texas at Dallas. His early work focused on the contribution of lead from gasoline to lead in blood. He continued working in the fields of toxicology and environmental medicine using stable-isotope dilution to measure concentrations of zinc in the brain and spinal cord and lead in blood serum and cerebrospinal fluid. He has used variations in lead-isotope ratios to determine how children acquire lead and exploited the long residence of lead in bone to study conditions, such as pregnancy, under which the skeleton might be resorbed. He has also been involved in studies in which animals have been dosed with stable lead isotopes, including investigations of lead poisoning from retained bullets and the mobilization of lead from the skeleton during pregnancy and after ovariectomy to stimulate menopause. He has also published on lead in the US diet and its unexpectedly high concentrations in cocoa and chocolate. He received his PhD in geochronology from the University of Witwatersrand, South Africa.

Jerry R. Miller is the Whitmire Professor of Environmental Science in the Department of Geosciences and Natural Resources at Western Carolina University. His research interests include assessment of the transport and fate of heavy metals in rivers and lakes, sediment geochemical fingerprinting with multivariate and isotopic methods, documentation of the effects of natural and anthropogenic disturbances on sediment yields and river processes, and restoration of riparian ecosystems. He received his PhD in geology from Southern Illinois University, Carbondale.

Peggy A. O'Day is professor of life and environmental sciences at the University of California, Merced. Dr. O'Day studies the chemistry, reaction, and transport of inorganic contaminants and species, primarily metal and metalloid elements, in surface and subsurface systems. She specializes in the use of spectroscopic and microscopic methods, especially synchrotron x-ray techniques, to determine element speciation and molecular-scale mechanisms of biogeochemical reactions in natural systems and laboratory analogues. She develops and applies thermodynamic, kinetic, and reactive transport models for synthesis and quantitative description of biogeochemical cycling, reactivity, transport, and bioavailability. She received a PhD in applied earth sciences from Stanford University.

John Toll is a partner at Windward Environmental LLC. He is recognized for his work in ecologic modeling, risk assessment, uncertainty analysis, and environmental decision analysis. He works primarily on complex contaminated sediment sites and leads teams that conduct remedial investigations and feasibility studies, ecologic risk assessments, and bioaccumulation modeling. He is the author or a co-author of numerous scientific papers on ecologic modeling and risk assessment and environmental decision analysis. Dr. Toll is a senior editor of the journal *Integrated Environmental Assessment and Management* and is a member of the board of directors of the Society of Environmental Toxicology and Chemistry (SETAC) North America. He earned his BS in chemical engineering from the University of Iowa and his PhD in engineering and public policy from Carnegie Mellon University.

Warren White is a researcher at the Crocker Nuclear Laboratory of the University of California, Davis. His research focuses on airborne particles and visibility impairment. He served on the review panel for the Air Quality Criteria for Particulate Matter Document and was a member of the US Environmental Protection Agency Clean Air Science Advisory Committee and its particulate-matter monitoring subcommittee. He has served on several National Research Council committees. He received his PhD in mathematics from the University of Wisconsin.

C

Open Session Meeting Agendas

COMMITTEE ON SOURCES OF LEAD CONTAMINATION AT OR NEAR SUPERFUND SITES

FIRST MEETING
Open Session: November 21, 2016
National Academies Keck Center, Room 100
500 Fifth Street, NW, Washington, DC 20001

PUBLIC AGENDA

1:00 **Purpose of Open Session and Introduction of Committee Members**
Edward Bouwer
Chair, Committee on Sources of Lead Contamination at or near Superfund Sites

1:15 **EPA Perspective on Study Objectives and Background**
Michele Burgess
Senior Biologist
Office of Land and Emergency Management
Office of Superfund Remediation and Technology Innovation
US Environmental Protection Agency

2:15 **Natural and Anthropogenic Sources of Lead in the Big River Watershed**
Teresa Bowers
President, Gradient
Michael Montgomery
Senior Attorney, The Doe Run Company

3:15 **Open Microphone – Opportunity for Public Comment**

4:00 ***End of Open Session***

SECOND MEETING
Open Sessions: January 18, 2017
Southeast Missouri Lead-Mining District

OPEN-SESSION AGENDA I: TOUR OF SOUTHEAST MISSOURI LEAD MINING DISTRICT

9:00 **Historical Perspective of Mining and Milling**
Art Hebrank, Site Administrator of Missouri Mines State Historic Site
Location: Missouri Mines State Historic Site, located southeast of the intersection of Highway 32 and Business 32/67 in Park Hills, MO

10:30 **Example of Residential Park Remediation**
Location: Haney Park, Located at the 500 Block of Crane Street (Near Intersection of Crane and Kelly Street), Park Hills, MO

11:00 **National Mine Tailings Site**
Location: Park Hills Industrial Park (near the intersection of Parkway Drive and S. Chamber Street), Park Hills, MO

11:45 ***Lunch Break***

1:00 **Mine Tailings Site Adjacent to Big River**
Location: Desloge Mine Tailings Site (Located between Landfill Road and W. Oak Street, Desloge, MO) and Bone Hole County Access Point (located off of Landfill Road, Desloge, MO)

2:00 **Example of State Park Remediation**
Bill Bonnell, Site Administrator
Location: St Joe State Park (located on Pimville Road about 2.4 miles southeast of the intersection of Highway 32 and Pimville Road from Park Hills, MO or about 4.25 miles northwest of the intersection of Bray Road and Pimville Road from Farmington, MO)

3:00 ***Tour Concludes***

OPEN-SESSION AGENDA II: PRESENTATIONS ON GEOLOGY AND MINING
Mineral Area College
North College Room, College Park
112 Dixie Kohn Drive, Park Hills, MO 63601

3:30 **Purpose of Open Session and Introduction of Committee Members**
Edward Bouwer
Chair, Committee on Sources of Lead Contamination at or near Superfund Sites

3:45 **Southeastern Missouri Geology and Mining in the Old and New Lead Belts**
Cheryl M. Seeger
Geologist
Missouri Geological Survey

4:30 **Mining, Milling, and Waste Handling Procedures at Active Lead-Mining Sites**
Mark Yingling
Vice President, Environmental, Health & Safety
Doe Run

5:15 ***Dinner Break***

OPEN-SESSION AGENDA III: PRESENTATIONS ON LEAD CONTAMINATION AND OPPORTUNITY FOR PUBLIC COMMENT
Mineral Area College
North College Room, College Park
112 Dixie Kohn Drive, Park Hills, MO 63601

6:45 **Purpose of Open Session and Introduction of Committee Members**
Edward Bouwer
Chair, Committee on Sources of Lead Contamination at or near Superfund Sites

7:00 **Environmental Transport and Deposition of Lead in Southeastern Missouri**
Emitt Witt
Hydrogeologist
US Geological Survey

7:45 **Lead Contamination Associated with Various Sources**
Valerie Wilder
Chief, Site Assessment Unit, Superfund Section, MO Department of Natural Resources
Hazardous Waste Program
Jonathan D. Garoutte
Chief, Bureau of Environmental Epidemiology, Department of Health and Senior Services

8:30 **Open Microphone – Opportunity for Public Comment**
NOTE: Those who want to make a public comment must register by 8 p.m. Each speaker has a maximum time limit of 5 minutes. Speakers are invited to submit relevant written materials.

9:30 ***End of Open Session***